CO_2 SEQUESTRATION TECHNOLOGIES FOR CLEAN ENERGY

Awareness and Capacity Building

The Editors

Dr. Syed Zahoor Qasim obtained his Ph.D. and D.Sc. degrees in Marine Science from the University of Wales, U.K. He has held positions of Director in several institutions and was Secretary to Government of India for 8 years; Vice-Chancellor, Jamia Millia Islamia and Member (Science) Planning Commission, Government of India. Besides being the author of 11 books and more than 300 scientific papers, he is an elected Fellow of all National Scientific Academies and a Fellow of the Third World Academy, Trieste (Italy). He is an Honorary Professor in four Indian Universities and has received Honorary Degrees of D.Sc. from four Universities including the Banaras Hindu University. He was the Leader of the First Indian Expedition to Antarctica in 1981-82 and built two Permanent Indian Stations (Dakshin Gangotri and Maitree) in Antarctica during his tenure as Secretary, Department of Ocean Development from 1981-88.

He was elected President of the National Academy of Sciences for the period 1983-84 and General President of Indian Science Congress for the year 1992-93. He was awarded Padma Shri in 1974 and Padma Bhushan in 1982. He was honoured by the "Lifetime Achievement Award" by the Oceanology International of U.K. in 1999, the first Ocean Science and Technology Award 2003-04 by the Government of India in 2003 and Gold Medal by the Asian Society in 2005. In 2008, he has received another Lifetime Achievement Award by the Prime Minister of India, Dr. Manmohan Singh at Visakhapatnam during 95th Session of the Indian Science Congress and SOFTI Biennial Award 2007 by the Hon'ble Chief Justice of India, Shri K.G. Balakrishnan at Cochin for his outstanding contribution in the field of Fisheries and Ocean Sciences. He also received Sir Syed Ahmad Khan International Award – 2009 on 17th October 2009 at Aligarh Muslim University, Aligarh. He is presently the Chairman, Centre for Ocean and Environmental Studies, New Delhi.

Dr. (Mrs.) Malti Goel received her Bachelor and Master's degrees in Physics from Birla Institute of Technology & Science (BITS), Pilani. Her two degrees in Solid State Physics including a Doctoral degree (1973) are from Indian Institute of Technology (IIT), Delhi. She is a recipient of gold medal in M.Sc. Physics (1967) from BITS, Pilani and a topper in D.I.I.T. Solid State Physics (1969) from IIT, Delhi. Prior to joining Government of India in 1982, she has been engaged in post doctoral research at IIT Delhi and was Research Associate in the Center for Materials Science. In the Department of Science & Technology (DST), Ministry of Science and Technology she has been heading the Science & Technology Advisory Committee Division. She served on several Programme Advisory Committees for promotion of thrust area programmes in Physical Sciences, Atmospheric Sciences, Greenhouse gas technology and Inter-sectoral energy related research programmes for capacity building on national scene. She has made significant contributions in energy policy research and has led carbon sequestration national research programme as Scientist 'G' and Adviser.

Dr. Malti Goel has 26 years of experience in science administration. She has been elected to represent India as Vice-chair to Technical Group of International Carbon Sequestration Leadership Forum (CSLF). She is recipient of Er. Abinash Chandra Chaturvedi award 2006 for excellence in Environment Science & Technology and became Fellow, National Environment Science Academy in 2008. She is adjunct professor Jamia Hamdard University and is on the Executive Board /Member of important scientific committees and professional bodies in the country. As Emeritus Scientist, Indian National Science Academy she organized ACBCCS-2009. At present she is CSIR Emeritus Scientist in the Center for Studies in Science Policy, Jawaharlal Nehru University, New Delhi.

CO_2 SEQUESTRATION TECHNOLOGIES FOR CLEAN ENERGY

Awareness and Capacity Building

— *Editors* —

Syed Zahoor Qasim

Malti Goel

2010

Daya Publishing House

Delhi - 110 035

ISBN 81-7035-660-1
ISBN 978-81-7035-660-8

Published by : **Daya Publishing House**
1123/74, Deva Ram Park
Tri Nagar, Delhi - 110 035
Phone: 27383999
Fax: (011) 23260116
e-mail : dayabooks@vsnl.com
website : www.dayabooks.com

Showroom : 4760-61/23, Ansari Road, Darya Ganj,
New Delhi - 110 002
Phone: 23245578, 23244987

Laser Typesetting : **Classic Computer Services**
Delhi - 110 035

Printed at : **Chawla Offset Printers**
Delhi - 110 052

PRINTED IN INDIA

Acknowledgements

The cooperation and support extended by the National Environment Science Academy and the facilities received from Indian National Science Academy in bringing out this publication are gratefully acknowledged.

Dr. (Mrs) Malti Goel

NATIONAL INSTITUTE OF ADVANCED STUDIES
Indian Institute of Science Campus, Bangalore 560 012, India

Prof. V.S. Ramamurthy
Director

Tele : 91+(0) 80-2360 1969 (O) Fax : 91+(0) 80-22185076
email : vsramamurthy@nias.iisc.ernet.in

Foreword

Development is a natural aspiration of every human being on the face of the earth. While a precise definition of development is yet to emerge, per-capita energy availability has been recognized as a critical factor for human development. Not only there exist wide disparities in per capita energy availability among populations in developed, developing and least developed countries, but the increasing global energy needs are also putting great pressure on the available energy resources. It is also known that increasing use of fossil fuels greatly increases the amount of carbon into atmosphere leading to unacceptable changes in the global climate. Apart from increasing the share of renewable energy resources in addressing the global energy demands, carbon capture and sequestrations have been recognized as potential mitigation measure.

I am happy that INSA organized a meeting on the awareness and capacity building in carbon capture and storage programmes. Many important topics including carbon capture and storage, current policy assessments, power sector carbon mitigation strategies, new coal combustion technologies and carbon capture materials research were discussed in this meeting. The need for open access and global partnerships in energy technology research were also highlighted in the meeting.

It is a great pleasure to give a Foreword to the book on Carbon Sequestration Technologies for Clean Energy. I congratulate Dr. (Mrs) Malti Goel for the initiative taken by her.

V.S. Ramamurthy
Director, NIAS, Bangalore
Former Secretary, Department of Science & Technology, Government of India

Preface

Global warming and climate change have become the most daunting environmental challenges for the mankind in 21st century. The climate issue has recently undergone heated debate in Copenhagen during the 15th meeting of Conference of Parties (COP) of World nations. As we are becoming more conscious of environmental degradation caused by increasing consumption of energy, each entity whether a source of energy or a household or an industry, it is compelled to take measures for reducing CO_2 emissions. In this context CO_2 sequestration has emerged as a technology option for reduction of its concentrations in the atmosphere. The technology to tackle this environmental catastrophe by carbon capture has given rise to the need for creating awareness on challenges in science & technology.

It is indeed a pleasure to introduce the book on *Carbon Sequestration Technologies for Clean Energy* about the processes by which carbon management takes place in the energy sector. The papers presented in the awareness and capacity building on carbon capture and storage program held during July 2009 addressed a range of topics in carbon capture and sequestration for clean energy production from fossil fuels are included. Important themes covered are; perspectives in carbon capture and sequestration technology, advancements in the pre-combustion, combustion and post combustion capture, bio sequestration, terrestrial sequestration and CO_2 storage in the oceans.

It is hoped that the book will be useful for the scientists, technologists, research students and to all others who feel interested about the subject. We are thankful to Shri Anil Mittal, Daya Publishing House for its timely publication and nice get-up. Despite the wide coverage of topics, there may be shortcomings and we welcome suggestions from our readers.

S.Z. Qasim *Malti Goel*

Prologue

Our life begin to end the day we become silent about things that matter

— From Archies 2010 in 'Stop Global Warming'

About a year ago I received the honor of joining Indian National Science Academy as Emeritus Scientist. Being a premier science academy in India, the offer was made in the Platinum Jubilee year to begin a Science Policy Studies cell at INSA for developing linkages between physical sciences with social sciences. While at INSA, it was my earnest desire to bring out a policy paper on emerging issues in Energy R&D and carbon sequestration.

I did get the privilege to organize the Awareness and Capacity Building on Carbon Capture and Storage (ACBCCS-2009) programme. It provided an introduction to challenges in addressing energy and climate change concerns related to excess CO_2 in the atmosphere. CO_2 sequestration - carbon capture and storage (CCS) is being looked as most effective way to mitigate CO_2 emissions in the atmosphere in the context of global warming. The meeting was inaugurated by Dr. V.S. Ramamurthy (Former Secretary, Department of Science & Technology). Dr. S.Z. Qasim (Former Member, Planning Commission) presided over the meeting and delivered a special address on Climate Change and Corporate Response. Shri R.V. Shahi (Former Secretary, Ministry of Power) delivered the keynote address.

The concept of carbon capture and storage still a nascent science, fascinates me. I recall my election to CSLF Technical Group to represent India as Vice Chair in its meeting held in New Delhi. I was to confront a great challenge to identify the national priorities related to science & technology in CCS while addressing the functions of CSLF under the guidance of its Policy Group represented by Shri R. V. Shahi from

India. I made sincere efforts to understand the phenomenon of CO_2 sequestration and to sensitize the energy industry in finding solutions through research & development. The *Technology*, which forms a milieu of economic development and progress in the society, is also relevant to address climate change concerns. Looking at the response from industry, the ACBCCS-2009 was a rewarding experience. It is gratifying that industry and academic institutions have supported it. I feel deeply indebted to Dr. S. Z. Qasim and Dr. V. S. Ramamurthy for their encouragement as chief patrons. I am grateful to Prof. N. K. Goel, Prof Nittala S. Sarma, Prof. Pratap Yadav, Dr M.O. Garg, Shri A. K. Mathur, Dr A. N. Goswami, and Shri S. D. Tripathi who honored me by their unstinted support. I am thankful to all those who have contributed to the success of ACBCCS-2009.

This book on **CO_2 Sequestration Technologies for Clean Energy** consists of lectures delivered in ACBCCS-2009. This is third in the series on *Global Warming*, a theme I have been pursuing since 1992. The first (Energy Sources and Global Warming, Allied Publishers 2005) has a note of appreciation from Dr A. P. J. Abdul Kalam then President of India who released it on 29th May 2006. The second book (Carbon Capture and Storage R&D: Technologies for Sustainable Future, Narosa 2008) is the outcome of the papers presented in an International Workshop held at Hyderabad. The book has chapters on post-combustion capture of CO_2 and underground geological sequestration by the national and international experts. The present book is about management of CO_2 by capturing it. Usefulness of various fixation approaches in the Indian context is discussed in detail. The technical themes included are,

— Carbon Capture Concepts, Issues and Perspectives

— Pre-combustion and Combustion Carbon Capture

— Post Combustion Carbon Capture

— Bio-sequestration,

— Ocean Sequestrations

— Sequestration in Coal Mines

— Role of Forests as Carbon Sinks

I would like to express my profound gratitude to Dr. Shailesh Naik, Secretary, Ministry of Earth Sciences and Prof. M. Vijayan, President Indian National Science Academy and T.R.C. Sinha, General Secretary National Environment Science Academy for the encouragement and support received for organizing ACBCCS-2009. I take this opportunity to thank Shri S.K. Sahni, Executive Secretary, INSA and Dr. V.S. Chintala Rao, Adviser, Ministry of Earth Sciences. The assistance from Dr. Kiran Arora and other colleagues at INSA is thankfully acknowledged.

My special thanks are due to guest speakers and invitees, who have submitted their lecture notes for the Proceedings. Although technical issues in diverse areas of CCS are covered in the book, policy dilemmas are still confronted. From the policy point of view India has vast environment policy framework which relies on its earlier policy Acts enunciated from time to time. The National Action Plan 2008 addresses the climate change issues. The CCS is not included as the cost of electricity generation becomes enormously high and coal becomes no longer affordable. The policy has rightly implied that technology is first developed and proven, and only after that it can be deployed. However, seeing the rapidity of global actions it may not be long that costs may come down and an integrated look on the various sources of energy and options of carbon fixation would become obligatory. A roadmap on aspects of technological feasibility and economic viability may have to be developed in view of India's energy dependence on coal. This is an area where India can also take an important position in the global scenario. A new thrust is therefore vital through policy guidelines and joint partnerships in R&D by involving different organizations from industry, academia and research laboratories.

Dr. (Mrs.) Malti Goel
Organizing Secretary, ACBCCS & Editor

Contents

Chapter 1

Corporate Response to Climate Change

S.Z. Qasim

Former Secretary, Department of Ocean Development and Member, Planning Commission; Chairman, National Environment Science Academy

In this address various symptoms of climate change are brought out which were highly varied since several centuries. In recent past human activities have played a havoc by overloading the atmosphere with CO_2.

Global Warming is caused when the atmospheric concentration of greenhouse gases like CO_2 increases. Excessive burning of fossil fuels like coal, petroleum is the main cause. The Greenhouse gases around the earth allow the sunlight to be trapped as heat and CO_2 does not escape. It is estimated that the annual average temperature by the end of the century will be 19°C from 14.43°C which was in the end of last century. This is followed by rise in sea levels which will increase to about 0.8m in 2100 from now, when it is 0.1m only at present. The CO_2 emission across the globe was 7 billions metric tons in the year 2000. It has been projected that by the year 2030, it would increase to more than 13 billion tons, when the water availability will get reduced by 30 per cent. The CO_2 concentration in the atmosphere was 379 ppm in 2005. Pre-industrial value of CO_2 was 280 ppm. By the year 2030, it is estimated that 25 per cent of the existing flora and fauna would become extinct. The greatest crisis would occur in marine fauna. Similar estimates indicate that if the temperature rise is half a degree centigrade, 17 per cent would be the fall in wheat production in India.

Presidential Address delivered in the Awareness and Capacity Building in Carbon Capture and Storage (ACBCCS-2009) Programme conducted from July 27-31, 2009 at Indian National Science Academy, Delhi.

India's initiatives to combat climate change have been praiseworthy because the share of renewable power to total power installed capacity is 7.75 per cent.

Corporate response will require a change in lifestyle. The various sources of CO_2 emission indicate that the highest is 15.7 per cent by the U.S., 7.9 per cent is by Europe and only 2.6 per cent by the Asia and Oceania. There is a proposal in the U.S. for constructing green buildings which conserve heat during the winter and allows enough cool airflow inside during the summer.

In 1997, an agreement was signed at Kyoto in Japan known as Kyoto Protocol. In this, it was agreed that the industrialized nations would cut down their emission containing CO_2 by about 5 per cent from 1990 level, during 2008 to 2012. The U.S. did not sign this protocol.

Formal talks on a post-2012 agreement on climate change began in Bali in December 2007. Since there is far too much money involved in climate change business for the G-8, they are likely to disregard it.

The former Vice President of U.S. Mr. A.A. Al Gore earned the Nobel Prize on his work on Climate Change. This Nobel Prize was shared by Intergovernmental Panel on Climate Change (IPCC) chaired by Dr. R.K. Pachauri of India. Significant points made are explained in the following presentation.

SYMPTOMS OF CLIMATE CHANGE

- ☆ Droughts and wildfire
- ☆ Floods and crop failure
- ☆ Destruction of climate – sensitive species
- ☆ Images of drowning polar bears
- ☆ Hurricane driven destruction of New Orleans
- ☆ Heat is continuing to rise
- ☆ The year 2006 was the hottest on record in the U.S.
- ☆ In Delhi the year 2020 will be the year without winter
- ☆ From January the temperature will soar to 35ºC
- ☆ Several food crops have failed and farmers are committing suicide by the dozen
- ☆ Since 1750, human activities have played a havoc by overloading the atmosphere with CO_2

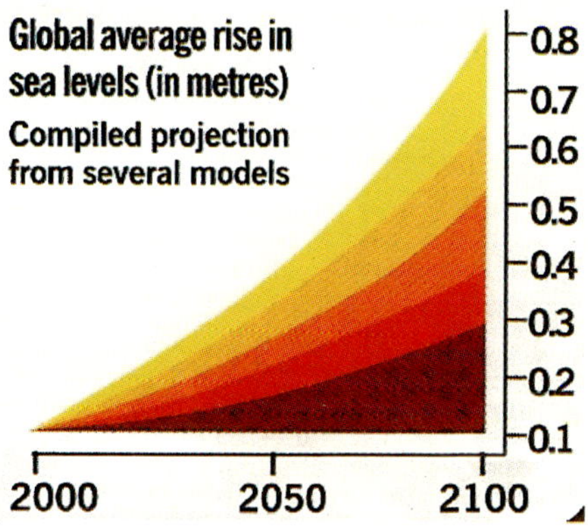

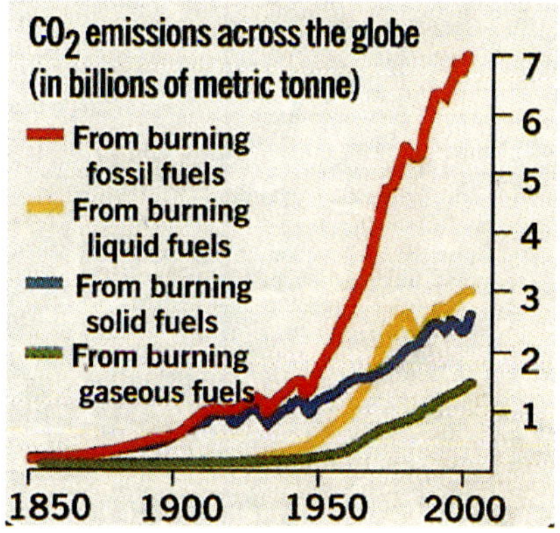

- ☆ 2030 is the year when per capita water availability will get reduced by 30% because of dryness
- ☆ 2035 is the reported year when the Himalayan glaciers may begin to disappear?

Source: IPCC

SOLAR PASSIVE BUILDINGS IN INDIA

State Bank of Patiala, Shimla

Solar Energy Centre, Gwal Pahari, Gurgaon

100 KW PV POWER PLANT AT KILTAN ISLAND, LAKSHADWEEP

CRISIS OF INCREASING GREENHOUSES GASES

379 p.p.m. in 2005

was the concentration of carbon dioxide (CO_2) in the atmosphere.

This is highest in 6,50,000 years. Pre-industrial value was 280 p.p.m.

By 2030

25% is the figure reported by several sources, of the existing fauna and flora likely to become extinct

5ºC

is the maximum expected rise in overall global temperature by the end of this Century.

Greatest Crisis

would occur in marine fauna because warmer ocean would lead to bleaching of coral reefs and shell dissolution of shelled marine animals.

Food Crisis

17% will be the reported fall in wheat production in India even if the temperature rise is half a degree centigrade

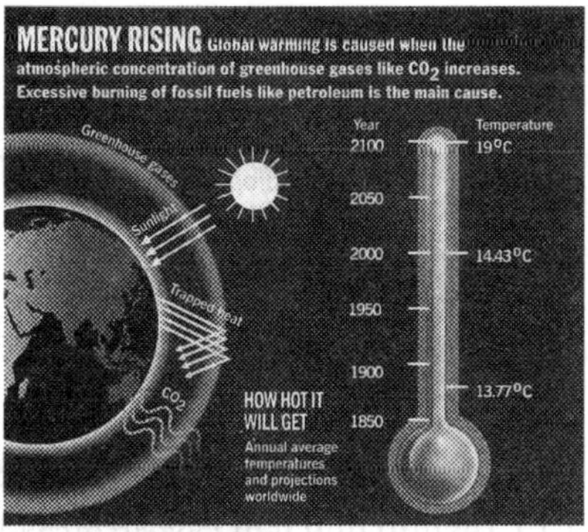

CORPORATE RESPONSE NEEDED

Changes in Life Style

- ☆ Use of solar devices to heat water.
- ☆ Use of Compact Flourescent Lamps.
- ☆ Construction of green buildings which will have insulation to retain heat and opening of windows in the direction of wind to replace air-conditioning.
- ☆ Computers should be turned off when not in use.
- ☆ All electrical gadgets should be turned off when not in use.
- ☆ Use of cars for short distances should be avoided.
- ☆ Air travel should be reduced if the work could be done by the use of telephone or e-mail.
- ☆ Switch to wind power is necessary — In India one company "Suzlon" earned $1.5 billion in wind turbine revenue in 2006.
- ☆ Plastic bags should be avoided.
- ☆ Use of paper should be cut down. The Indian habit of putting everything down on paper should be avoided. Use of e-mail and phone can save thousands of trees.
- ☆ Public transport should be made use of as far as possible.
- ☆ Power-generating plants based on fossil fuel will have to modify their ways.
- ☆ We should switch to biofuels - ethanol made from corn or sugarcane.
- ☆ All industries have to demand clean technology. Developed countries should sell clean technology to developing countries at a cheaper rate.

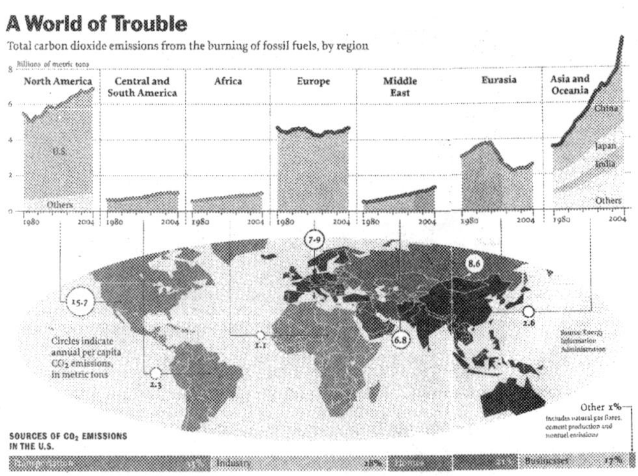

Between 1990 and 2004 energy consumption rose 37% in India and 53% in China.
Americans per capita emission of CO_2 is about 21.75 tons.
In China it is 4.03 tons and India 1.12 tons.

INDIA'S INITIATIVES TO COMBAT CLIMATE CHANGE

- ☆ Total Installed Power Capacity in India = 1,32,110.21 MW
- ☆ Total Renewable Power = 10251.68 MW
- ☆ Per cent of Renewable Power to Total Power = 7.75%
- ☆ Total Installed Wind Power Capacity in the World = 73,904 MW as on 31.12.2006
- ☆ India is 4th after Germany (20,622 MW), Spain (11615 MW) and USA (11,603 MW)

RENEWABLE ENERGY TECHNOLOGIES – ACHIEVEMENTS (As on 31.3.2007)

Biogas Plants (nos.) - 3.9 million

Solar PV - 75 MW

a) Street Lights - 61,321

b) Home Lighting Systems - 3,13,859

c) Lanterns - 5,65,658

d) Power Plants - 1.87 MW

Solar Hot Water Systems Installed in Pune

Solar Water Heating - 1.65 million sq. m.

Wind Power - 7092.00 MW

Small Hydro - 1975.60 MW

Bagasse Cogen - 615.83 MW

Biomass Power - 524.80 MW

Waste to Energy - 43.45 MW

Wind farm in Rajasthan

8.4 MW Wind farm at Chitradurga, Karnataka

A 50,000 litre per day (LPD) Solar Water Heating System installed at Chelsea Jeaus Textile Company, Manesar, Gurgaon

AL GORE: WE ARE FACING PLANETARY EMERGENCY

"Change is possible. We know that, and we can alter our relationship with the world around us to create a sustainable future.....

The environment is much more than a policy position to me; it is a profoundly moral obligation.....

We only have one earth. And if we do not keep it healthy and safe, every other gift we leave for our children will be meaningless."

- ☆ We are facing what is probably a planetary emergency. It is difficult to see the reality of this emergency because it's outside the boundaries of any previous experience in human history. But there are three factors that have combined to completely and utterly transform the relationship between the human species and the earth's ecology.
- ☆ The first is the population explosion, which has quadrupled the human population in the last 100 years.....in the course of a single life time — the generation born after World War II — we have moved from two billion to over nine billion.
- ☆ Second, our technology's power has magnified a thousand folds in just the last century. When this new power has been used without adequate wisdom and when the new power is multiplied by six-and-a-half billion people, what we get is a very new relationship between humankind and the planet.
- ☆ The third factor is a psychology, a philosophy, an attitude — a way of living on the earth — which is dominated by a focus on the short term to the exclusion of the long term. The atmospheric shell of the planet is thin and we are now changing its composition — disrupting the climatic pattern that existed since the last Ice Age 11,000 years ago — But we can still change it — we have the time, technology, everything except the political will. Though in a democracy, political will is a renewable resource.
- ☆ Leading scientists in the world are openly saying we probably have less than 10 years in which to make dramatic change in the pattern that now represents business as usual.
- ☆ To reduce the accumulation of global warming pollutants like carbon dioxide, methane and a few other gases.

Contd...

AL GORE: WE ARE FACING PLANETARY EMERGENCY

Contd...

Q. The US is not a signatory to the Kyoto Protocol. What can the world do about it?

"There is a technical distinction here – the US did sign the protocol. I persuaded President Clinton to sign it before our administration left office. But in order for the US to be party to the treaty, it has to be ratified by the Senate…

I believe that 15 minutes after President Bush leaves office, the US will have a new policy on global warming, regardless of which political party wins the next presidential elections".

Q. What advice would you give India on global warming, especially as it embarks on its economic success story?

"India faces an exciting new set of opportunities in the 21st century. But it is crucial that India make an accurate accounting of the true cost of energy sources and environmental policy.......

It should not get mesmerised by short term calculations..India and China are emerging as the dominant economies of the world in the later part of this century. But as the largest democracy, India has unparalleled opportunity to lead the world to sustainable development. I am encouraged by the many positive signs I see here".

NOBEL FOR CLIMATE PANEL LED BY INDIA'S PACHAURI

"New Delhi/Oslo: In a major boost to the international campaign for action against global warming and an honour for India, the UN climate panel, headed by Indian environmental warrior Rajendra Kumar Pachauri, was awarded the prestigious Nobel peace prize on Friday."

- ☆ The Intergovernmental Panel on Climate Change (IPCC) shares the award with former US vice president and climate campaigner Al Gore, the Norwegian Nobel Committee announced in Oslo.
- ☆ IPCC, a UN body comprising 3,000-odd atmospheric scientists, oceanographers, ice specialists, economists and other experts, is the world's top scientific authority on global warming and its impact on the planet.
- ☆ The Nobel committee cited the recipients for their work "to build up and disseminate greater knowledge about manmade climate change, and to lay the foundations for measures that are needed to counteract such change".
- ☆ The award is an oblique recognition of the work done by 67-year-old Pachauri, who was elected IPCC chairman in April 2002, and under whose leadership the UN panel grabbed the world's attention on the urgent need to fight climate change.
- ☆ IPCC's fourth report, published this year, gave the starkest view yet on global warming, warning that climate change was already on the march and that rising temperatures fuelled the risk of drought, flooding and violent storms.
- ☆ Reacting to the honour, Pachauri told TOI, "It is a positive development also because the prize today has brought climate on everyone's radar. The key findings of IPCC will get more attention. It will prompt governments to take decisions."
- ☆ Addressing reporters a few hours before the announcement, Pachauri had said: "I have no expectation; never had any. Things today for me are like any other day." But the scene at New Delhi's The Energy and Resources Institute, which Pachauri has headed as director-general for decades, soon became one of loud celebration. He was thronged by excited colleagues and journalists. Amid the cheering, the bearded climate crusader said he was "privileged" that IPCC was sharing the award with Al Gore."

KYOTO PROTOCOL AND BEYOND

In 1997 an agreement was signed at Kyoto in Japan known as Kyoto Protocol.

- ☆ In this it was agreed that the industrialised nations would cut down their emission containing CO_2 that is warming the atmosphere disrupting weather patterns and thus leading to climate change.
- ☆ The reduction would be about 5% relative to 1990, starting from 1998 through to 2012 level.
- ☆ U.S. did not sign this protocol.
- ☆ The countries included were G-8 namely U.S., Canada, U.K., France, Germany, Italy, Japan and Russia.
- ☆ Five developing countries namely India, China, South Africa, Brazil and Mexico also participated in the Kyoto Meeting.
- ☆ Formal talks on a post-2012 agreement on climate change began in Bali in December 2007.
- ☆ Since there is far too much money involved in climate change business for the G-8, they are likely to disregard it.

BALI MEETING

- ☆ As a follow up of Kyoto Protocol, a meeting on Climate Change was held in Bali, Indonesia on 6th December 2007 in which American negotiators participated.
- ☆ As we are aware, the U.S. is world's largest producer of such gases which are responsible for Global Warming.
- ☆ The Kyoto agreement expires in 2012. In Bali, it was decided that they will introduce 70% pollution cut by 2050.
- ☆ U.S. did not sign but was agreeable to implement the necessary recommendations.
- ☆ The most important thing that happened in Bali is that Australia which was a non-signatory to the Kyoto Protocol has signed in Bali.

Chapter 2

Carbon Capture and Storage Technology: A Possible Long Term Solution to Climate Change Challenge

R.V. Shahi

CMD, Energyinfratech Ltd. and Former Secretary, Ministry of Power, Govt. of India

Industrial revolution in the world required massive exploitation and utilisation of energy resources. Much before the discovery of petroleum fuels, civilisation and industrialisation grew around fossil fuels as the most important of the energy resources. It is because of these reasons that utilisation of coal emerged as the most important component of the energy consumption. That is why when we look at the carbon emissions in developed countries, we find the per capita emission of CO_2 disproportionately higher in these countries as compared to the emissions in undeveloped and developing economies. The pace of growth continued. Desire and aspirations of people in poor countries to have their living standards enhanced, also grew. Communication and media explosion enabled people living in poor countries to understand the enormity of differences in the styles of living of people in developed countries vis-à-vis their own conditions. Public pressures on political systems led to different countries evolving strategies to charter a rapid economic growth. This needed massive doses of energy. Countries like India and China, more particularly China,

---

Delivered as Keynote Address in the Awareness and Capacity Building in Carbon Capture and Storage (ACBCCS-2009) programme conducted from July 27-31, 2009 at Indian National Science Academy, Delhi.

depended heavily on fossil fuels as these fuels were more cost-effective. The absolute quantum of CO_2 started increasing at a much more rapid pace in developing economies than in industrialised nations. This was inevitable, obviously not unusual, because people in these countries badly needed to increase their energy consumption levels.

Environmental ramifications of consumptions of fossil fuels, notably among them being coal, have always been known. Adverse impact of greenhouse gases has been studied and disseminated over last two to three decades. But in the last ten years, in the wake of economic growth of large developing countries, carbon emission has been noticed with concern by all the countries in the world. Studies initiated under the purview of Inter-governmental Panel on Climate Change have particularly brought into focus the implications of ever increasing CO_2 emissions. In spite of countries having committed to comply with the obligations under the Kyoto Protocol of 1997, the implementation have been far from being satisfactory. Instead of reducing the emission levels as compared to the Base Year 1990 in almost all cases, emissions increased in the countries which are highly industrialised and which committed to reduce CO_2 emissions.

Several initiatives, therefore, were needed to address this situation. One such initiative under the leadership of the U.S.A. Carbon Sequestration Leadership Forum (CSLF) was launched in June, 2003 with a number of nations signing the CSLF Charter in Washington on June 25, 2003. Sequestration means separation and storage. I had the opportunity and privilege of being associated directly with the drafting and finalisation of this Charter. I recall that in the last week of June when the representatives of 19 countries, including India, were to meet and finalise the CSLF Charter to be signed by the Ministers of different countries, an initial draft had been forwarded by the U.S.A. to all the countries. In India there were differing perceptions and opinions among various Ministries and Departments whether or not India should be signatory to this Charter. Without naming the Ministries and Departments, the views differed in the following manner – (a) U.S.A., which has not even ratified the Kyoto Protocol, wants to divert the attention of the world through this new CSLF initiative, (b) Seriousness on the part of U.S.A. is less, publicity aspect is more, (c) India has a long way to go in increasing its energy production base and larger amount of CO_2 emission is inevitable. Its per capita CO_2 emission is so low that initiative like Carbon Capture and Storage is irrelevant to India, (d) The cost of such technologies would be so excessive that it would be unbearable for Indian economy, (e) Developed countries, once we become a signatory, may try to push these technologies however unaffordable they may be, to countries like India, which we can ill-afford.

While we had views on these lines, we also had different perspectives by others. They opined that climate change issues were so important that many countries would go along with such initiatives. If India keeps itself out, it may not serve much purpose. It could be looked upon by others that it is not serious about such important issues like greenhouse gases and climate change concerns. U.S.A. will, in any case, push through this initiative. There is no harm in participating in this initiative so long as deployment of technologies, if they are not cost-effective, are not binding on us. This became the corner stone of India's approach in this matter. Accordingly, we concluded that India should participate in the Charter but should protect its position while finalising the draft of this Charter.

When the official level discussion took place, in Washington on June 24, 2003, to finalise the draft of the Charter, I participated in the core drafting team. The original draft included the formulation "To facilitate the development and deployment of improved cost-effective technologies............" We successfully argued that it would not be desirable to include in the draft the phrase "Deployment". Around this core concept, the entire draft was restructured and alterations were made in different paragraphs for a harmonious reading. Given below are a few important extracts from the Charter. These relate to – (i) Purpose of the CSLF, (ii) Functions of the CSLF, (iii) Organisation of the CSLF, (iv) Research and Intellectual Property Right.

"1. Purpose of the CSLF

To facilitate the development of improved cost-effective technologies for the separation and capture of carbon dioxide for its transport and long-term safe storage; to make these technologies broadly available internationally; and to identify and address wider issues relating to carbon capture and storage. This could include promoting the appropriate technical, political, and regulatory environments for the development of such technology.

2. Function of the CSLF

The CSLF will seek to:

2.1: Identify key obstacles to achieving improved technological capacity

2.2: Identify potential areas of multilateral collaborations on carbon separation, capture, transport and storage technologies

2.3: Foster collaborative research, development, and demonstration (RD&D) projects reflecting Members' priorities

2.4: Identify potential issues relating to the treatment of intellectual property

2.5: Establish guidelines for the collaborations and reporting of their results

2.6: Assess regularly the progress of collaborative R&D projects and make recommendations on the direction of such projects

2.7: Establish and regularly assess an inventory of the potential areas of needed research

2.8: Organize collaboration with all sectors of the international research community, including industry, academia, government and non-government organizations; the CSLF is also intended to complement ongoing international cooperation in this area

2.9: Develop strategies to address issues of public perception

2.10: Conduct such other activities to advance achievement of the CSLF's purpose as the Members may determine

3. Organization of the CSLF

3.1: A Policy Group and a Technical Group will be formed. Unless otherwise determined by consensus of the Members, each Member will make up to two appointments to the Policy Group and up to two appointments to the Technical Group. Other individuals may attend the Policy Group and Technical Group meetings as deemed necessary by the appointed representatives.

3.2: The Policy Group will govern the overall framework and policies of the CSLF, periodically review the program of collaborative projects, and provide direction to the Secretariat. The Group should meet at least once a year, at times and places to be determined by its appointed representatives. All decisions of the Group will be made by consensus of the Members.

3.3: The Technical Group will report to the Policy Group. The Technical Group will meet as often as necessary to review the progress of collaborative projects, identify promising directions for the research, and make recommendations to the Policy Group on needed actions.

3.4: The CSLF will meet at such times and places as determined by the Policy Group.

3.5: The principal coordinator of the CSLF's communications and activities will be the CSLF Secretariat. The Secretariat will: (1) organize the meetings of the CSLF and its sub-groups, (2) arrange special activities such as teleconferences and workshops, (3) receive and forward new membership requests to the Policy Group, (4) coordinate communications with regard to CSLF activities and their status, (5) act as a clearing house of information for the CSLF, (6) maintain procedures for key functions that are approved by the Policy Group, and (7) perform such other tasks as the Policy Group directs. The focus of the Secretariat will be administrative. The Secretariat will not act on matters of substance except as specifically instructed by the Policy Group.

3.6: The Secretariat may, as required, use the services of personnel employed by the Members and made available to the Secretariat. Unless otherwise agreed, such personnel will be remunerated by their respective employers and will remain subject to their employers' conditions of employment.

3.7: The U.S. Department of Energy will act as the CSLF Secretariat unless otherwise decided by consensus of the Members.

3.8: Each Member will individually determine the nature of its participation in the CSLF activities.

4. Open Research and Intellectual Property

4.1: To the extent practicable, the R&D fostered by the CSLF should be open and non-proprietary.

4.2: The protection and allocation of intellectual property, and the treatment of proprietary information, generated in R&D collaborations under CSLF auspices will be defined by implementing arrangements."

Based on this Charter, a Policy Group was set up as the Apex Policy Organisation. I had the privilege of being a Member on this Policy Group till the time I relinquished the charge of Secretary Power in January, 2007. Regular interactions in the Policy Group led to better understanding of the issues. However, not much progress could be made because of complete lack of clarity on financing of Research Projects. I Co-Chaired a Group on Financing Issues and made a Presentation to the entire CSLF Group. Suggested options are outlined in the Box.

Suggested Options

- ☆ Preferable option is to create a separate fund for promotion of CSLF projects in developing countries with suitable contribution from developed country members.
- ☆ Example of US indicating commitment of USD 50 million under Methane to Market Partnership.
- ☆ Fund size may be modest to begin with (USD 100 million).
- ☆ Contribution may be on the basis of a reasonable criteria.
- ☆ Per Capita CO_2 emission could be a basis.

Per Capita CO_2 Emission

Country	*Tons of CO_2*
United States	19.66
Australia	17.36
Canada	16.93
Netherlands	11.02
Russia	10.43
Germany	10.15
Denmark	9.52
Korea	9.48
Japan	9.47
United Kingdom	8.94
Italy	7.47
Norway	7.28
South Africa	6.65
France	6.16
Mexico	3.64
China	2.57
Brazil	1.77
Colombia	1.26
India	0.97

World Average– 3.89 Tones

(Source: Key World Statistics (2004) by International Energy Agency.)

Suggested Contributions

☆ U.S.A	USD 30 million
☆ Other Developed Countries	USD 5 to 10 million

In the Policy Group, it was decided to set up Technical Committees which would identify projects to be undertaken by different countries. India's presence was duly recognised by election of Malti Goel, Senior Scientist from Department of Science and Technology, Govt. of India with India becoming the Co-chair on the CSLF Technical Group. A list of projects, which were indentified, is given below. We also organised a few scientific conferences and workshops on this subject in India.

"CSLF RECOGNIZED PROJECTS

(*i*) ARC Enhanced Coal-Bed Methane Recovery Project

The ARC Enhanced Coal-Bed Methane Recovery Project is a pilot-scale project (3 test wells) located in Alberta to evaluate a previously developed process of CO_2 Injection into deep coal beds for simultaneous sequestration of the CO_2 and liberation (and subsequent capture) of coal-bed methane.

(*ii*) CANMET Energy Technology Centre (CETC) R&D Oxyfuel Combustion for CO_2 Capture (PDF)

A pilot-scale project (0.3 megawatt-thermal) located near Ottawa, Ontario, that will demonstrate oxyfuel combustion technology with CO_2 capture. The goal of the project is to develop energy-efficient integrated multi-pollutant control, waste management and CO_2 capture technologies for combustion-based applications and to provide information for the scale-up, design and operation of large scale industrial and utility plants based on the oxy fuel concept.

(*iii*) CASTOR

The CASTOR "CO_2 from Capture to Storage", is a European initiative grouping 30 partners (industries, research institutes, and universities) representing 11 European countries, including CSLF members France and Norway. The project is partially funded by another CSLF member, the European Commission, under the 6th Framework Program. CASTOR`s overall goal is to develop and validate, in public/private partnerships, all the innovative technologies needed to capture and store CO_2 in a reliable and safe way.

(*iv*) CO_2 Capture Project

This is a pilot-scale project that will continue the development of new technologies to reduce the cost of CO_2 separation, capture, and geologic storage from combustion sources such as turbines, heaters and boilers. CCP is an international public private R&D partnership.

(*v*) CO_2 GeoNet

This focus of this project, which began in 2004, is on geologic storage options for CO_2 as a greenhouse gas mitigation option, and to assemble an authoritative body for Europe on geologic sequestration. Major objectives include formation of a partnership consisting, at first, of 13 key European research centers and other expert collaborators in the area of geological storage of CO_2, and identification of knowledge gaps in the long-term geologic storage of CO_2 and formulation of new research projects and tools to eliminate these gaps.

(vi) CO_2 Separation from Pressurized Gas Stream

This is a small-scale project to evaluate processes and economics for CO_2 separation from pressurized gas streams. Testing will utilize membranes developed in Japan at a test facility near Pittsburgh, Pennsylvania, United States. The proposed project, which began in 2003 and scheduled for completion in 2006, will evaluate primary promising new membranes under atmospheric pressure. The next stage is to improve the performance of the membranes for CO_2 removal from the fuel gas product of coal gasification and other gas streams under high pressure.

(vii) CO_2 SINK

This is a pilot-scale project that will test and evaluate CO_2 capture and storage at an existing natural gas storage facility near Berlin, Germany, and in a deeper land-based saline aquifer. A key part of the project will be monitoring the migration characteristics of the CO_2 stored.

(viii) CO_2 STORE

This large-scale project is a follow-on to the current Sleipner project, which involves injection of about one million metric tons per year of CO_2 into an offshore geologic saline formation beneath the North Sea. This next phase will involve continuation of monitoring of the field to track CO_2 migration (involving a seismic survey) and additional studies to gain further knowledge of geochemistry and dissolution processes. There will also be several preliminary feasibility studies for additional geologic settings (in Wales, Germany, Denmark, and Norway) of future candidate project sites. The goal of the project is to develop sound scientific-based methodologies for the assessment, planning, and long-term monitoring of underground CO_2 storage, both onshore and offshore.

(ix) Feasibility of Demonstration of Capture, Injection and Geologic Sequestration of CO_2 in Basalt Formations of India

This is a pilot-scale project that will develop the feasibility to demonstrate the viability for deep bed injection of CO_2 in the sedimentary rocks underlying India's very widespread basalt formations. Sites selected for the project will be basalt covered areas with minimum trap thickness of 600 meters of underlain sedimentary rocks. Intensive investigations to discover the fate of the CO_2 using a broad range of geo-physical and geo-chemical techniques and the development of numerical models and leakage risk assessments will be taken up. Knowledge gained from this project will be applicable in other parts of the world.

(x) Development of China's Coal Bed Methane Technology/Carbon Dioxide Sequestration Project

This is a pilot-scale project, begun in 2002, which will evaluate reservoir properties of selected coal seams of the Qinshui Basin of eastern China and carry out field testing at relatively low CO_2 injection rates. This project will address unique issues related to the storage capacities in coal beds and coal bed methane rate enhancement and will provide information on the coal reservoirs including gas adsorption/desorption characteristics, injectivity variations, permeability changes and CO_2 storage capacities. The main objective of this project is to use the results from the Alberta, Canada Enhanced Coal-Bed Methane project and make improved estimates on the technology's performance, cost and benefits, first in south Qinshui basin and then in other coal basins in China.

(*xi*) ENCAP

The ENCAP Project is a broad-based collaboration between industry, authorities, universities and institutes, with a goal of developing new pre-combustion CO_2 capture technologies and processes for power generation.

Activities within the project are being focused in six main areas:

- ☆ Process and Power Systems
- ☆ Pre-Combustion Decarbonisation Technologies
- ☆ O_2/CO_2 combustion (Oxy fuel) Boiler technologies
- ☆ Chemical Looping Combustion
- ☆ High-Temperature Oxygen Generation for Power Cycles and
- ☆ Novel Pre-Combustion Capture Concepts

The results from ENCAP will enable power companies to launch a new design project by the end of the decade which could lead to a large demonstration plant, with potential for additional wide commercial exploitation, by the year 2020.

(*xii*) Frio Project

The Frio Project is a pilot-scale project located near Houston, Texas, that will demonstrate CO_2 sequestration in an on-shore underground saline reservoir. The project involves injecting relatively small quantities of CO_2 into the reservoir and monitoring its movement in the formation for several years thereafter.

(*xiii*) Geologic CO_2 Storage Assurance at In Salah, Algeria

"In Salah Gas" is a commercial joint venture which recovers natural gas containing up to 10% CO_2 from several geological reservoirs in Algeria for processing and delivery to markets in Europe. As a result of the processing, 1 million metric tons of CO_2 per year is produced, this is re-injected into carboniferous sandstone reservoirs at a depth of 1,800 metres. The In Salah project, which is expected to run for five years, is an industrial scale demonstration of CO_2 geological storage. The goals of the project are to verify that long-term assurance of geologic storage can be met by short-term monitoring and to demonstrate that such projects represent a viable greenhouse gas mitigation option. An additional objective is to set precedents for the regulation and verification of the safe, secure and cost-effective geological storage of CO_2 in a gas reservoir.

(*xiv*) ITC CO_2 Capture with Chemical Solvents

This is a pilot-scale project (4 metric tons per day CO_2 capture) located on a flue gas slip stream of a lignite-fueled power plant near Regina, Saskatchewan, Canada, that will demonstrate CO_2 capture using chemical solvents. Supporting activities include bench- and lab-scale units that will be used to optimize the entire process using improved solvents and contactors, develop fundamental knowledge of solvent stability and minimize energy usage requirements. More than $5 million has so far been spent on construction of the pilot facility at the project site and another $3 million on a pilot plant (1 metric ton per day CO_2 capture) at the University of Regina where additional testing is taking place. The goal of the project is to develop improved cost-effective technologies for separation and capture of CO_2 from flue gas. Current research is demonstrating significantly reduced regeneration energy requirements.

(xv) Regional Carbon Sequestration Partnerships

The Regional Carbon Sequestration Partnerships project, which began in September 2003, is a broad-based collaboration of industry and the research community to help identify and test the most promising opportunities for implementing sequestration technologies in the United States and Canada. The overall project structure is comprised of more than 240 organizations from the United States and six other CSLF Members within an area that includes 40 American states, 4 Canadian Provinces, and 3 Native American Nations. Results from Phase I of the Partnerships have identified more than 600 gigatons of storage capacity in domestic geologic formations, including saline reservoirs, depleted oil and gas fields, coal seams, shale, and basalt formations. Additional characterization during Phase II is expected to identify more potential sinks and refine the estimates determined during Phase I.

(*xvi*) Regional Opportunities for CO_2 Capture and Storage in China

This project is a broad-based collaboration between United States and Chinese organizations that is being coordinated by U.S./China Energy and Environmental Technology Center. A multinational team will compile key characteristics of large man-made CO_2 sources, including power plants, iron and steel facilities, cement kilns and refineries. The team will also study candidate geologic storage formations that exist across China and develop estimates of geologic CO_2 storage capacities. A proven methodology previously developed under the auspices of the IEA Greenhouse Gas R&D Programme will be modified as appropriate and applied to assess the distribution of CO_2 storage potential within China, and the infrastructure needs and costs associated with the capture of CO_2. The goal of the project is to characterize the technical and economic potential of CO_2 capture and storage in China, and assess the ability of geologic CO_2 storage to help mitigate current and future CO_2 emissions there.

(*xvii*) Weyburn II CO_2 Storage Project

This is a commercial-scale project that will utilize CO_2 for enhanced oil recovery at a Canadian oil field. The goal of the project is to determine the performance and undertake a thorough risk assessment of CO_2 storage in conjunction with its use in enhanced oil recovery."

Technical content of each of the CSLF recognized project was quite high. The subject of deployment was raised in few other instances but was not agreed to. Subsequently, when the U.S.A. proposed for "FutureGen Zero Emission" project, India not only participated, but in fact, became the first country to collaborate with U.S.A. with some monetary participation. In the year 2006 I signed, on behalf of the Govt. of India, the Agreement on "FutureGen Zero Emission Project" along with my counterpart in the U.S. Energy Department. I was also on the Steering Committee of this project. India's commitment on the Research and Development is adequately reflected by its active participation in all these initiatives.

My own belief on this subject has been that unless various countries commit substantial research and development funds, network various components of research projects, it is unlikely that cost-effective technologies will be available to the developing countries, so that they can deploy these technologies in the power generation process. An example in case could be the IGCC. This project has been talked for over three decades, but it has not come to any visible size and shape in countries like India. Carbon Capture and Storage can, no doubt, be a technology of the future for which every country should put its effort. But, to think that their deployment is going to be easily acceptable in the near future, would be a too simplistic assumption. This should, however, not discourage or dampen our efforts to participate in research and development related to CCS. It may not be a cost-effective technology in the short or medium term, but it may definitely emerge as a possible long term solution to the challenge of CO_2 emissions and climate change. We must recognise that, despite our thrust on development of all our hydroelectric potential, plans to augment substantially the nuclear power capacity, commitment to harness all possible other renewable energy sources (wind, bio-mass, solar etc.), since our power demand is going to grow at the annual rate of 8 to 9% over next 25 years, heavy dependence on coal will have to continue. Therefore, our power development strategy must explore and find solutions to the increasing CO_2 emission problem. CCS Technology will obviously have to be researched intensively to make it cost-effective. This is indeed a challenge to scientists and technologists.

Chapter 3

Perspectives in CO_2 Sequestration Technology and an Awareness Programme

Malti Goel

Former Adviser, Ministry of Science and Technology, and CSIR Emeritus Scientist, Jawaharlal Nehru University, New Delhi, India

SUMMARY

This paper begins with the genesis of global warming and past evidences of CO_2-temperature coupling. Presently, fossil fuel combustion is contributing a major share to global CO_2 emissions in the atmosphere. In view of India's dependence on coal as dominant source of energy in thermal power generation, the CO_2 sequestration technology expects to address two most critical problems of 21st century namely; energy security and climate change. It is foreseen that fossil fuels will continue to be the major source of energy in the coming decades and CO_2 capture and fixation or utilization becomes desirable for the energy security of coal dominant nations. CO_2 sequestration can provide increase in biomass, enhanced fertilization and value addition through production of fuels. This in turn would help in mitigation of climate change.

The CO_2 capture technology in fossil fuel based energy production can be introduced at three stages *viz.*, pre-combustion, combustion and post combustion. An overview of the technology perspectives in these for CO_2 stabilization from its point sources and various possibilities of storing it away from the atmosphere safely are described. Pathways to CO_2 sequestration comprise bio-sequestration, terrestrial and underground trapping, storage in oceans as well as recovery of energy fuels. International perspectives in this emerging energy technology and India's policy responses as well as National Action Plan on Climate Change are discussed. For creating awareness on scientific and technological challenges the programme on Awareness and Capacity Building in Carbon Capture and Storage (ACBCCS-2009) is organized.

INTRODUCTION

Imagination is more important than knowledge, for knowledge is limited, whereas imagination embraces the entire world–stimulating progress, giving birth to evolution

— *Albert Einstein*

Environment, energy and education are fundamental to human growth and development. Education helps in understanding the environment and many other concepts that revolve around it including that of sustainability. In the present day context CO_2 concentrations in the atmosphere, having most significant share in greenhouse gas emissions, are obstructing the sustainable growth. While energy is essential for the industrial growth and economic well being, it also causes environmental pollution due to fossil fuel consumption and coal based energy generation. Increasing energy consumption is depleting our limited mineral resources and proving intimidating to mankind due to global warming phenomenon. The environmental consequences of industrial growth in one part of the world and future projections of growth in other parts are being determined through the devastating climate change predictions such as failure of agricultural crops, health hazards, water scarcity and degradation of the ecosystem as a whole. Although energy and economic development are inseparables, it was never envisaged that growth would lead to such consequences.

Looking Back and Looking Forward

Mankind has been in a state of turmoil ever since prehistoric era and the consumption of energy has been rising gradually [1]. However, disequilibrium in consumption pattern of energy resources came about after the industrial revolution, which began with the invention of steam engine and provided mobility to man. Energy in its present form has proved handy to human race, but its indiscriminate use has created enormous problems in the society. The energy consumption has grown rapidly with the exploration of fossil fuel reserves; availability of huge reserves of coal led to low cost electricity generation and finding of cheap oil resources resulted in mammoth increase in transportation. Nobel laureate (1995) Paul J. Crutzen postulated that since the beginning of the 19th century mankind has entered into Anthropocene era [2]. He said "Industrial output increased 40 times in the past century and energy use 16 times, water 9 times; and almost 50 per cent of the land surface has been transformed by the society in the 20th century". Several other studies have analyzed the future energy growth patterns and have concluded that the current pattern of growth will lead to unilateral increase in greenhouse gas concentrations in the atmosphere.

Since the prehistoric era, the changing CO_2 concentrations have been associated with earth's temperature variations, which caused either Ice age or global warming. The CO_2-Temperature coupling was demonstrated about 300 million years ago. The oxygen isotopic and stomatal index data supported it [3]. Uplift of Himalayas 45 million years ago and increased upwelling of dissolved nutrients in south west Pacific Ocean led to sequestering high CO_2 present on earth in the Eocene era. In the nearer past, over a million of years ago there is further evidence that the temperature changes match with CO_2 concentrations measured over Antarctica during the last Glacial period. The record of atmospheric CO_2 concentrations during past 1, 60,000 years [4] obtained from deep ice cores drilled at 2,083m Antarctica ice cap at Russian Vostok base suggests a phased relationship between CO_2 and global mean surface temperature (Figure 1).

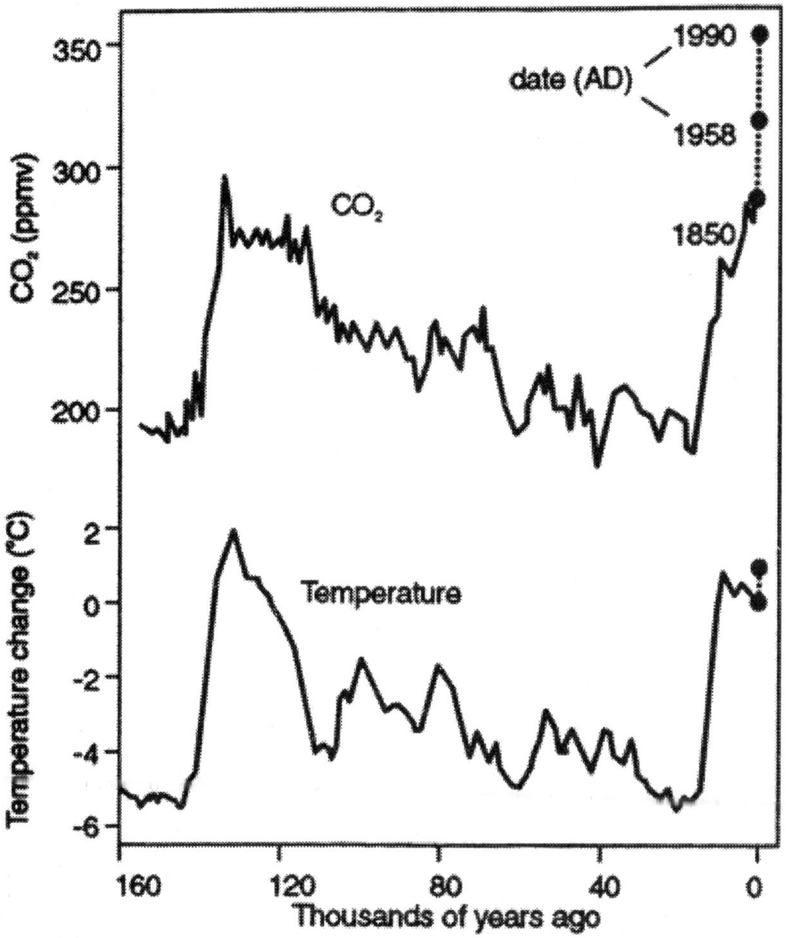

Figure 1: Changes in Global Atmospheric CO_2 and Temperature Since Last 160,000 Years (Source-IPCC)

Scientific View of Global Warming

In 1896, Arrhenius [5] attributed increased fossil fuel combustion to global warming due to accumulation of CO_2 in the atmosphere and said that temperature may rise up to 5ºC in future. Many studies in favour and countering this theory were put forward in the subsequent years. In the 1960s and early 1970s, a cooling trend was also experienced in the Northern hemisphere. It led many scientists to predict that the planet was entering a new 'little ice age'. This fear later proved short lived as the warming trend again became significant from 1980s onwards.

Carbon dioxide amount measured at Mauna Loa has indicated a steady rise in the atmosphere since 1950s as shown by Keeling Curve (Figure 2). The explanation of global warming is based on physical principles of solar radiation interactions with the earth system. In terms of energy the average solar radiation falling on outer rim of the earth is measured as 1368W/m^2 per year. Out of this, only 342W/m^2 is received on the earth surface due to presence of atmosphere. The earth surface reflects back 102 W/m^2 to space. The trace gases present in the atmosphere absorb part of the Earth's radiated energy, giving rise to natural greenhouse effect. This process is contributing to maintain average earth surface temperature at around 288K. Global warming is thus the enhanced greenhouse effect causing further warming of the earth's surface and its lower atmosphere due to accumulation of anthropogenic greenhouse gases *viz.* carbon dioxide (CO_2) as well as other major and minor greenhouse gases namely; nitrous oxide, methane and chloro-florocarbons, perflourocarbon, tropospheric ozone and sulphur hexafluoride. Among these CO_2 has largest contribution.

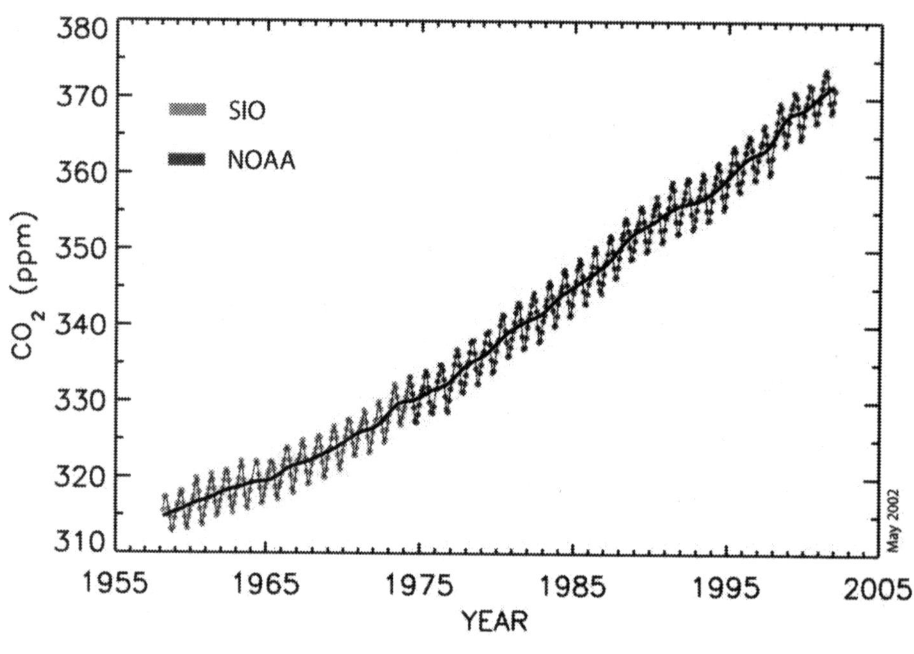

Figure 2: Keeling Curve on CO_2 Concentrations Observed at Mauna Loa

Scientific assessments of Inter Government Panel on Climate Change (IPCC) and its various reports have established consensus on occurrence of global warming. The average global CO_2 concentrations are seen to rise to 380 parts per million (ppm) (2005 data) against the acceptable average of 280 ppm and any rise above 450 ppm would be detrimental. The fourth IPCC report (AR4) has indicated that the global temperature has risen by 0.74°C in the last 100 years (1906 to 2005) and the projected temperature rise up to the year 2100 can be anywhere between 2.4 to 6°C, whereas sea level may rise between 19cm to 59cm. Climate change has become the most daunting challenge of the 21st century requiring urgent actions. This is due to the disastrous impacts of global warming on Planet earth and its climate namely; extreme weather events, more frequent disasters, increase in the floods and droughts, decline in crop productivity, water scarcity and vector borne diseases among others. In the ultimate analysis we may conclude that air pollution and its exchanges in Earth system have led to severe distortion in the climate pattern and could destroy the human race, if no action is taken.

Use of Fossil Fuels

The fossil fuels are meeting about 80 per cent of the total energy needs. Coal is dominant source of energy having 40 per cent share in world electricity, and it is also main source of air pollution and greenhouse gas emissions. Globally it is estimated that 4.3 billion tonnes of coal was burned in 2005 to meet energy needs of society. The other sources of energy such as; hydro, nuclear, Sun, wind, biomass, ocean and geothermal (renewable resources) meet the 20 per cent of world energy need and contribute more to solid waste as well as liquid effluents and less to air pollution. However, each energy technology has environmental implications in some form or other and it requires intensive analysis and more research to achieve sustainable progress. What Milan Kendara said in the context of life '*Everything turns problematic, questionable, subject of analysis and doubt ......*' is also true for energy technologies in solving energy-environment nexus.

All countries are differently endowed with resources available for energy generation. In some countries electricity is coal dominant, in some others it is hydro power dominated or nuclear power dominated. Per capita generation/consumption and hence per capita CO_2 emissions also varies from country to country. For selected countries CO_2 per capita is shown in Figure 3. In India, coal is having 69 per cent share in total electricity generation and 70 per cent of energy needs of manufacturing and processing industries are met through coal. The share of different fuels in electricity generation is shown in Figure 4. Per capita emissions in India are low, about 1.3 ton per annum as against world average of four tons. Energy security for meeting the basic needs of people is leading to a concern for clean environment and sustainable energy future. The dependence on coal is expected to rise and requirement has been projected [6] to become anywhere between 1600Mt and 2000Mt per annum (with and without renewables) by the year 2031.

CO_2 CAPTURE AND STORAGE

We need to have a technology for CO_2 sequestration or carbon capture and storage (CCS) for stabilizing greenhouse gas concentrations in the atmosphere. The CCS technology has three major technologies *viz.* capturing of CO_2 in the atmosphere emitted from large point sources; fixing it or transporting it to a possible location where it can be safely stored; and finally process of fixation as depicted in Figure 5. Point sources are considered as the most feasible for CO_2 capture like natural gas reservoirs, coal based power plants and energy intensive industries.

The captured CO_2 can be converted into liquid form, making its transportation easier to a desirable location for disposal. It can also be converted into useful products. The CO_2 in liquid form is injected into depleted oil and gas reservoirs; underground rocks-which slowly react to mineralization; deep saline formations, below the ground or under the sea bed. In addition to these, CO_2 can be injected for enhanced oil recovery from depleting oil reservoirs and enhanced coal bed methane recovery from abandoned coal mines or un-mineable coal seams.

TECHNOLOGY PERSPECTIVES IN CO_2 FIXATION

Carbon cycle in the earth system occurs from exchanges of carbon dioxide in the atmosphere with the hydrosphere, biosphere and lithosphere. In Nature, the process of CO_2 fixation occurs in many ways as follows.

(*i*) CO_2 dissolves in water turning into HCO_3

(*ii*) It forms $CaCO_3$ by reacting with minerals

(*iii*) Plants capture CO_2, convert it into organic matter in the presence of day light, a process known as photoautotroph or photosynthesis

(*iv*) Algae capture CO_2 to form bio diesel

(*v*) Bacteria and Achaea convert it into organic matter, using energy derived from oxidations of molecules

(*vi*) Oceans absorb CO_2 and contribute to growth of phytoplankton

Through these exchanges earth maintains a natural carbon balance. As the concentration of CO_2 in the atmosphere is increasing, natural carbon removal system is disturbed and global warming is being caused. For sequestration of excess concentration of CO_2 various technologies are invoked as discussed below.

Clean Coal Technology

In coal based thermal power generation the CO_2 sequestration technology can be used either pre-combustion or during combustion or post combustion. Pre-combustion capture takes place before the power is generated. Coal is first converted into Syn gas or liquid fuels. The Syn gas comprises mainly of carbon monoxide (CO) and hydrogen (H_2). The CO can be converted into CO_2 by using a shift reactor and can be separated. Hydrogen (H_2) can be used for pollution free power generation. Coal gasification plants using Integrated Gasification Combined Cycle (IGCC) system have higher efficiencies and combined with CCS they are expected to be near zero emission plants in future. Other advanced clean coal technologies namely, coal bed methane exploration and underground coal gasification also offer to reduce CO_2 emissions in the atmosphere during production and use of coal based energy [10].

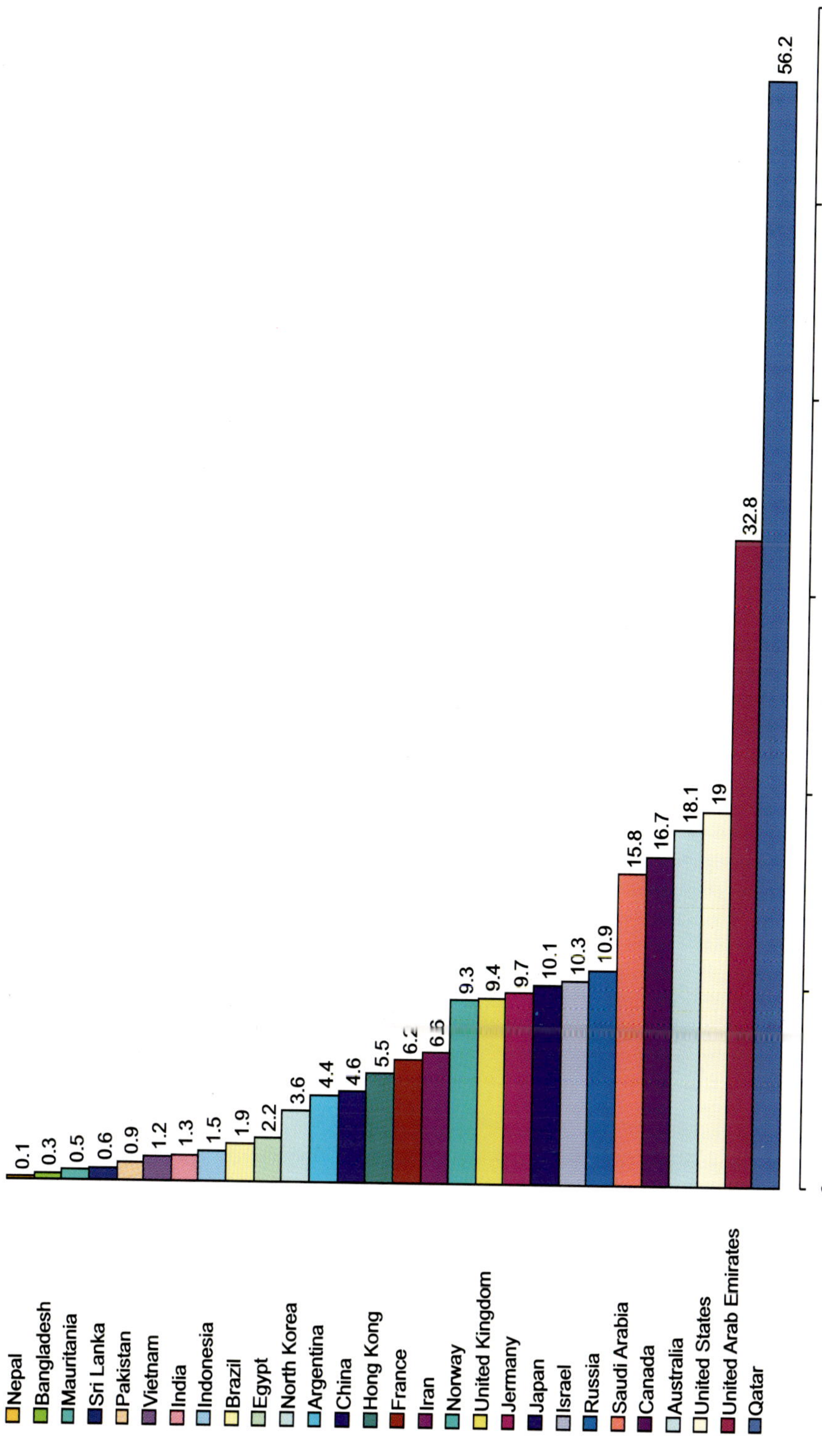

Figure 3: Per Capita CO_2 Emissions in Selected Countries of the World in 2006 [7]

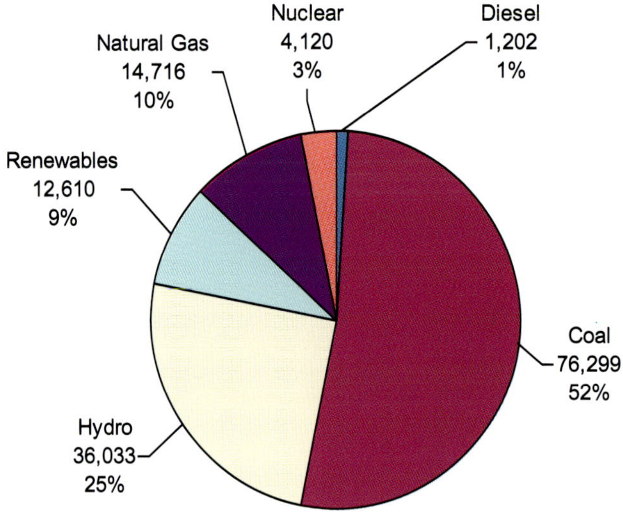

Figure 4: Share of Different Sources of Energy in Electricity Generation (2007) in India [8]

During combustion process there are two possibilities; (*i*) using supercritical and ultra-supercritical coal combustion efficiency can be increased, CO_2 emissions per unit of generation reduced (*ii*) using technologies like oxy fuel combustion and

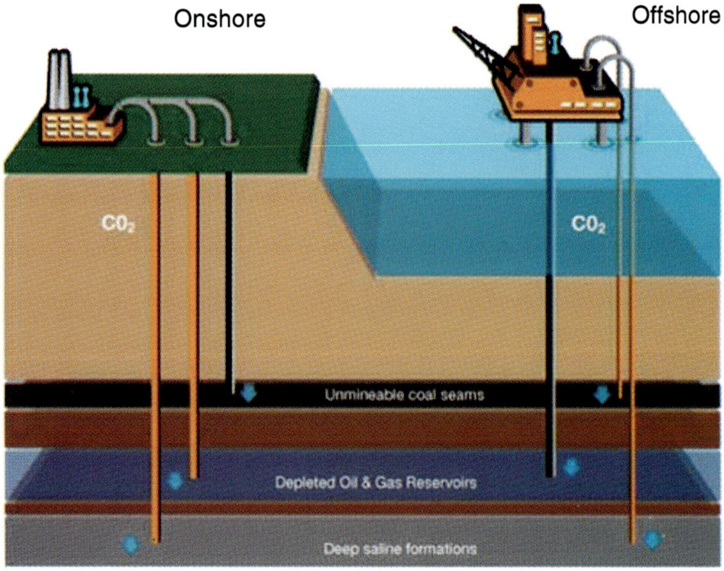

Figure 5: Different Options of Carbon Management through Carbon Capture and Storage [9]

chemical looping, CO_2 concentrations in flue gas are increased and could be easily removed. A steam cycle with steam pressure above 226 bars and temperature above 537°C is called supercritical. Compared to sub critical coal combustion, the expected increase in efficiency is by 5 per cent. The technology of ultra-super critical combustion operating at 357 bars/625°C is also under development. It may enhance the efficiency further. In the second option of oxy fuel combustion, pure oxygen is used in place of air during the combustion process. The outcome is steam energy and 90 per cent CO_2 is the flue gas. The CO_2 separation cost is considerably reduced and capture can become more cost-effective.

Post combustion capture is the third stage alternative, in which CO_2 is separated from the flue gas and is stored away from the atmosphere or utilized in industrial activities. The flue gas from a coal fired power plant has 8-14 per cent CO_2 by volume as depicted in Table 1.

Table 1: Flue Gas Composition of a Coal Fired and Gas Fired Thermal Power Plant

Pollutant	*Coal Fired*	*Gas*
CO_2	8–14 (per cent)	4 (per cent)
Oxygen	0.5 (per cent)	–
Nitrogen	81 (per cent)	81 (per cent)
SOx	300–3000 (ppm)	1 (ppm)
NOx	100–1000 (ppm)	100–500 (ppm)
Particulates	1000-10000 (ppm)	10 (mg/m^3)

At this percentage number of tests have been conducted using amines for CO_2 scrubbing and technique is fully developed. However, their application to large-scale operation leads to almost doubling the cost of electricity by introducing significant energy penalty and additional process cost. Substantial progress is being made in developing other processes like use of polymeric membranes for carbon capture, physical adsorbents and using nanotubes to separate greenhouse gases from smoke stacks. Notwithstanding these developments, there are many areas of clean coal technology in which further research is required to be done [11]. The CO_2 sequestration technologies for coal use are summarized below;

- Chemical absorption/solvent post combustion
- Physical adsorption methods
- Pressure swing and electrical swing adsorption techniques
- Cryogenic fractionation
- Gas separation membranes
- IGCC/Oxy fuel combustion
- Chemical looping
- Underground Coal Gasification
- Coal bed methane production

CCS and Industrial Energy

Among the various sectors, the industry sector consumes about 40 per cent of the total energy generated worldwide. Energy intensive industries such as iron, steel, cement, aluminum and oil processing etc. have an important role in economic growth and infrastructure development of a nation. These industries contribute most to CO_2 concentrations and are facing greatest challenge in reducing greenhouse gas emissions. In EU countries research for finding zero emission technology or minimizing CO_2 emissions per unit of steel production has been intensified.

The CO_2 reduction is possible on two way, (i) by using less energy for producing the same goods or, (ii) by producing goods which consume less energy during use. By applying energy efficient and energy conserving technology 25-40 per cent energy could be saved. Many such initiatives have been taken as summerized in [12] and research for finding zero emission technology or minimizing CO_2 emissions per unit of production are under investigation. Metal slag from industry is proving good absorber of CO_2. In India distribution of greenhouse gas emissions from various industries in 2005 is shown in Figure 6. Bureau of Energy Efficiency has introduced measures to increase energy efficiency of products and appliances. In post Kyoto Regime new policies for reducing sectoral emissions such as PAT (Perform, Achieve and Trade) are proposed.

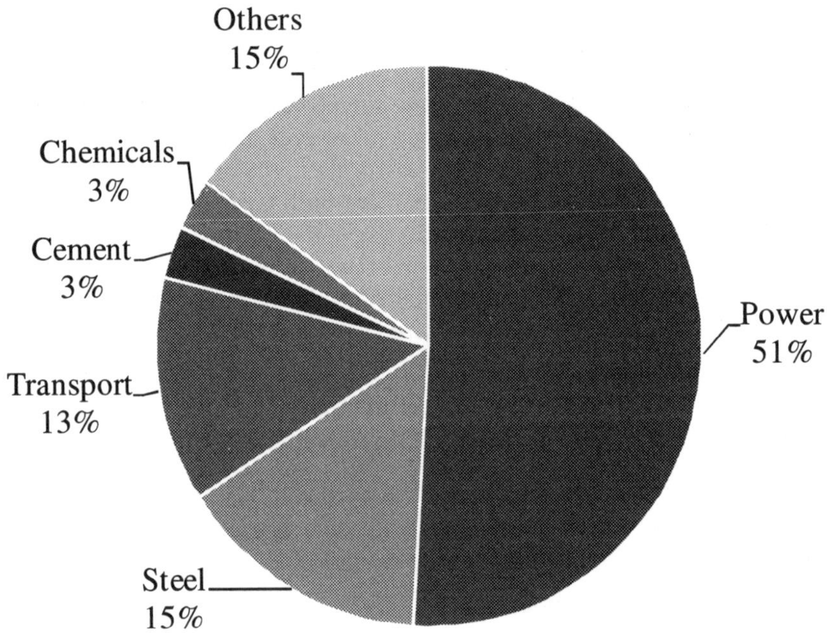

Figure 6: Greenhouse Gas Emissions from Different Industries in India, 2006

Underground CO_2 Trapping

Both active and passive underground trapping of CO_2 are being envisaged. The idea to cage CO_2 in natural geological surroundings for its storage has taken shape as a means to permanently removing it from the atmosphere. Both active and passive underground trapping of CO_2 are envisaged. It can be done by burying it in deep saline aquifers and in rocks and minerals as discussed in [13]. Rapid developments are taking place in evaluation of results of CO_2 storage in the underground deep aquifers and in the sedimentary basins under the sea bed. Sleipner, Norway has achieved tremendous success in this. It has injected 1 million tons of CO_2 every year in saline aquifers under the sea bed, since 1996.

To understand the reaction taking place between CO_2 and mineral rocks, active storage in basaltic rocks is being pursued for better integrity for CO_2 storage. Calcium and magnesium silicates react to form carbonates minerals. A process of geochemical trapping can effectively eliminate risk of CO_2 leakage. Regarding the safety of underground CO_2 storage, future studies on the size of the bore hole, post injection behavior of rocks, soil or water and geochemical analysis will provide the some required answers. Each geological setting is different and underground CO_2 trapping studies would need more detailed geomorphology studies. Long-term seismic studies will also have to be done. There is also a need to develop geo modeling and methodologies for long-term tracking of CO_2 injected.

Energy Fuels by CO_2 Sequestration

The process of underground storage of CO_2 and consequent changes in the viscosity of fluids can provide additional fuel for energy. The CO_2 will mix with the crude in the depleting oil reservoirs and will help flush out remaining reserves by changing the oil properties and making it flow easily. At the surface the gas mixture containing both CO_2 and natural petroleum gas will have to be separated from the oil before it is sent for refining. The CO_2 can be stripped out and petroleum gas sold or alternatively the full mixture can be compressed and recycled back into the oil fields to extract more oil. CO_2 management through this approach will be a highly challenging task [14]. It will have to be adopted when all the conventional methods of enhanced oil recovery have failed. A CO_2-EOR project designed to minimize CO_2 emissions back to atmosphere with appropriate incentives would have an important role in producing more energy and assuring energy security.

Like oil fields, unmineable coal seams can also prove to be potential reservoir for CO_2 storage. On average three molecules of CO_2 are absorbed for displacing one molecule of methane (CH_4). By injecting CO_2 in coal seams, coal bed methane recovery can be enhanced. More investigations in the geology of coal fields and coal properties are essential to be carried out for understanding these phenomena and are discussed in detail in the later chapters.

Land Sequestration: A Ground Truth

Terrestrial or land sequestration in plantations occurs through the well known process of photosynthesis. Advance crop species and cultivation practice could be designed to increase the uptake of CO_2 through enhanced photosynthesis rate. Bio sequestration processes offer opportunity for CO_2 to be recycled for value addition and fuel production. Micro mediated CO_2 sequestration using algae and carbonic anhydrate enzyme catalysis are other emerging options. Advances in genomic sciences are providing new ways of CO_2 fixation.

Earth's land surface, which is a natural reservoir of CO_2 in the soil as well as in forests can sequester CO_2 in soil and biota. In the soil CO_2 respiration occurs with the decomposition of waste, roots and other organic matter. Olivins in the soil as well as the wetlands are great CO_2 absorbers. Reclamation of waste lands together with afforestation has potential for CO_2 sequestration for both below and above the ground, while it may also add to growing carbon markets. There are many challenges in it, which shall from the part of discussion in the following chapters.

Storage of CO_2 in Oceans and Iron Fertilization

Oceans are mammoth reservoirs of CO_2 and are potential candidate for CO_2 sequestration in the sea water at different depths. It appears much simpler to disperse liquid CO_2 into oceans. Dispersal at shallow depths of less than 300m may however release it back to the atmosphere, through surface plumes. Injecting it to a depth of 1000m or so is likely to delay the process atmospheric release. But this may endanger the survival of marine species. Liquid CO_2 injected at a depth of 3000m would confine it to form a permanent lake, being denser than water. This option is likely to avoid any harm to marine ecology, but the cost of storage in this case will be enormously high. Physico–chemical mechanisms for disposal of frozen CO_2 in thermohyline zones are also under study [15]. CO_2 fixation in marine cynobacteria–a microbiological process, appears another attractive option.

Another technique for safety of marine ecology is to use iron filings on the upper surface of Oceans. CO_2 catalyzes production of phytoplankton and marine food. Large-scale experiments are being undertaken to test the efficacy of iron filings catalyzed ocean fertilization by CO_2 sequestration in different marine zones. The most recent one is Indo-German LOHAFEX (LOHA-Iron, FEX-Fertilization Experiment) conducted in early 2009. Ocean sequestration is bound by international regulations and Laws of seas and there are associated technical issues as well, which are being addressed.

POLICY PERSPECTIVES–INTERNATIONAL

The technologies of CO_2 fixation are highly multi-disciplinary and their implementation would offer not only research and development opportunities but also require a number of cross-cutting issues to be addressed. In the international policy perspective the issues connected with increasing greenhouse gas emissions and their consequences on environment are addressed by the Intergovernmental Panel on Climate Change (IPCC) formed in 1988. The IPCC in its endeavor to develop a consensus on global warming issues has produced a series of documents, as scientific assessment reports and has deliberated on the climate change concerns from various predictions and observations. In its fourth report (AR4) released in 2007, it is stated that global warming is irrevocable and policy measures were urgently required to mitigate climate change. In this context IPCC Special Report on Carbon

Capture and Storage [16] discussed various options and has evaluated them. An assessment of global point sources of CO_2 with emissions of more than 0.1 Mt of CO_2 per year has been made. According to this estimate emissions from 4492 major power sources added to 10.5 Bt of CO_2 per year and 2953 energy intensive industries added to 3.5 Bt of CO_2 in 2005. In the concluding report it has been suggested out that there are gaps in currently available knowledge regarding some aspects of CCS.

According to IEA study [17], the increasing knowledge and experience would reduce the uncertainties and thus facilitate decision-making with respect to CCS technology for climate change mitigation. The Stern Review report on the Economics of Climate Change (2006) has been another important policy document. It determines the effects of global climate change and global warming on the world economy. Investment of 1 per cent of gross domestic product (GDP) of a country could solve the climate crisis. Early action has been emphasized since the cost of handling accelerated climate change may increase further.

Whether CO_2 sequestration makes a credible climate change mitigation option or not has been debated in several international forums. Many policy and technology related options are being put forward for stabilization of CO_2 concentration in the atmosphere. In the wedge analysis [18], among its several wedges of actions to be taken, carbon capture and storage is suggested to be one of them. Considering the enormity of the problem the G8 Group of countries have evolved an international programme to develop and deploy CCS in as many coal power stations as possible. The G8 group proposes to explore possible standards for CCS by supporting 20 demonstration projects on CCS by 2020. Global Carbon Capture and Storage Institute has been setup in Australia in 2009. It is contemplating to facilitate initiation of research collaborations in CCS projects. Science Academies from G8+5 countries have also expressed the need to pursue the development, demonstration and deployment of economically efficient and technologically safe CCS, and to explore the establishment of standards.

While these efforts to develop a suitable technology for CCS may continue to grow, individually or in cooperation with others, it is imperative that each government should make its own policy and each household/industry takes steps to reduce CO_2 emissions, without waiting for an international agreement or for any external help. Each country has to make an intensive and all out efforts in this direction since the weather is becoming warmer, climate is becoming unpredictable and the cost of keeping it under check increasing every minute.

INDIA'S POLICY RESPONSE

It is important to mention here that India has a long standing policy on environment protection starting from Water Act of 1974, onwards to the National Action Plan on Climate Change announced in 2008. National Environment Policy 2006 first recognized that anthropogenic emissions from fossil fuel use, certain agricultural and industrial activities as well as deforestation were mainly responsible for climate change. It suggested the need for abatement, as it may cause catastrophic disruptions of livelihoods, human health and could result in significant economic costs to the nation.

India as signatory to Kyoto Protocol, a mechanism to stabilize greenhouse gas concentration ratified by countries participating in United Nations Framework Convention on Climate Change (UNFCCC), has no commitment to reach any specific targets in emission reduction. Yet India's policy pertaining to the production and use of energy by way of promotion of energy efficiency; appropriate mix of fuels and primary energy sources, including nuclear, hydro and renewable sources; abatement of pollution; increasing forestation; growth of mass transport, besides differentially higher growth rates of less energy intensive services sectors as compared to manufacturing has resulted in a relatively GHGs benign growth and sustainable development path. The emphasis has been on the need for adaptation to future climate change and to greater participation of Indian industry in Clean Development Mechanism (CDM). The CDM is a market mechanism, which depends on global emission reductions, for which CERs (Certified Emission Reductions) priced at 8-10 Euro per ton of CO_2 are earned by developing countries (over a period of 10 years), whereas the credit goes to developed countries.

India is also a founder member of Carbon Sequestration Leadership Forum (CSLF), a programme organized by the Department of Energy, USA in 2003 to engage in collaborative R&D. The author had privilege to attend as a member of Indian delegation and as Vice-Chairperson to CSLF Technical Group to its various meetings. It was emphasized during the discussions that the clean coal technology is the solution for carbon capture and storage in the developing countries to minimize CO_2 emissions while continuing to grow, and these should be made accessible. These views were heard and appreciated.

In India research on carbon sequestration began with the support from industry and Government of India [19]. Inter-sectoral interaction meets were held with the stakeholder's participation to develop thrust areas of research. A National Programme on CO_2 Sequestration (NPCS) research has been initiated under the umbrella of the Department of Science and Technology, Government of India in 2006, from the perspective of basic and applied research with the participation from academic institutions and R&D laboratories across the country. Thrust areas under for the programme are identified as; (i) CO_2 Sequestration through Micro-algae Bio-fixation Technique (ii) Carbon Capture Process Development (iii) Terrestrial Agro-forestry Sequestration Modeling Network (iv) Policy development studies [20]. Feasibility and exploratory studies on CO_2 sequestration in saline aquifers, basalt rocks and in depleted oil fields for enhanced oil recovery have also been started. Areas of research on ultra super critical and oxy fuel coal combustion technologies, and geo-modeling and monitoring tools and capabilities were proposed.

To give thrust to technology development India participated in FutureGen, world's first coal based zero emission plant and also in one of the CSLF projects. In 2006 India joined the Asia Pacific Partnership in Clean Development and Climate (AP6). The author had opportunity to learn from some of its workshops and also participated in the Workshop on Short-term Opportunities in Coal Sector for Carbon Capture and Storage, organized under the umbrella of Subsidiary Body on Scientific and Technology Actions, UNFCCC and in the preparatory meeting for global institute.

The National Action Plan on Climate Change (NAPCC) 2008 has devised a strategy to address global climate change in the following eight area:

- ✰ Increasing the Share of Solar Energy
- ✰ Implementing Energy Efficiency measures on large scale.
- ✰ Launching Sustainable Habitats
- ✰ Managing Water Resources effectively
- ✰ Safeguarding Himalayan Glaciers and Ecosystem.
- ✰ Enhancing Eco-System services
- ✰ Making Agriculture responsive to climate change
- ✰ Setting Strategic knowledge mission

The action plan encompasses a very broad and extensive range of measures and proposes to address the global climate change issues relating to Food Security, Water Security, Eco-system and Energy Security among others. The deployment of CCS has not been agreed due to high cost and high risk. It is implied that technology should be developed cost-effectively and proven for permanency of storage of CO_2, and only after that the question of deployment should be raised. The technology at present is in research phase and in India stress is on initiation of CCS research.

ACBCCS- 2009

ACBCCS is acronym for Awareness and Capacity Building in Carbon Capture and Storage. A programme on the subject is organized in Delhi from 27-31 July, 2009. In the conference on ACBCCS, 14 participants attended as nominee of from major stakeholder industries and academia. Eminent experts and scientists from organizations across the country delivered lectures for creating awareness on CCS technology [21]. This book contains important addresses, invited talks, guest lectures and scientific papers presented in the ACBCCS-2009 to cover a range of topics in carbon capture and sequestration for clean energy production from fossil fuels. The focus has been on advancements made in the international scene, policy perspectives and multi-disciplinary research being done in India. On the final day of the five day programme, an open roundtable discussion on 'Should carbon be priced?, on the regulatory and policy related issues like; Should cap-and-trade approach is right approach to be adopted?, Should CCS become acceptable under Clean Development Mechanism?, What should be the integrated approach in terms of R&D, technology transfer, regulation etc. is held and forms part of the deliberations.

CONCLUSIONS

CO_2 sequestration technology expects to address two most critical problems of 21st century namely; energy security and climate change. It is foreseen that fossil fuels will continue to be the major source of energy in the coming decades and CO_2 capture and fixation or utilization becomes desirable for producing clean energy and thus energy security of coal dominant nations. CO_2 sequestration can also provide increase in biomass, enhanced fertilization and value addition through production of fuels.

This in turn would help in mitigation of climate change. While international and national policies are targeting to arrive at some agreement on climate change issues at Copenhagen, we hope that a strategy for sustainable development around CCS evolves and clean energy technology development is promoted. This would help in increasing support to research and development in CCS and integration of climate change into the development process. In the long run, costs may come down and an integrated look on the various sources of energy and options of carbon fixation would be required.

ACKNOWLEDGEMENTS

ACBCCS-2009 is supported by Ministry of Earth Sciences, Government of India and National Environment Science Academy. The author is grateful to Prof M. Vijayan, President, Indian National Science Academy (INSA) and Dr V. S. Ramamurthy, Former Secretary DST for the encouragement. She is indebted to Dr. T.N. Hajela, eminent economist and educationist, Former Joint Secretary, University Grants Commission for the guidance and his very helpful suggestions.

REFERENCES

[1] Simmons I. G., 2008, Global Environmental History, Edinburgh University Press.

[2] Crutzen Paul J., 2008, An example of geo engineering cooling down Earth's climate by sulphur emissions in the stratosphere, in Predictability in Science: Accuracy and limitations, Eds. W. Arber, N. Casbibbo, M. S. Sorondo, Pontificia Academia Scientiarvm Actam, 19, 83-95.

[3] Retallack G. J., 2002, Phil. Trans. R. Soc. Lond. A, 360, pp 659-673.

[4] www.ipcc.ch/pdf/assessment-report/ar4/

[5] Arrhenius Svante, 1896, On the Influence of Carbonic Acid in the Air Upon the Temperature of the Ground, Philosophical Magazine 41: 237-76.

[6] Integrated Energy Policy, 2008, Report of the Expert Committee, Planning Commission 2006, Government of India and Ministry of Coal, Govt of India, Annual Report.

[7] U.S. Energy Information Administration, Independent Statistics and Analysis, EIA, 2009, Energy Outlook, www.eia.doe.gov

[8] All India Electricity Statistics, General Review, 2008, CEA, Min. of Power, Govt of India.

[9] World Coal Institute (www.worldcoal.org)

[10] Goel Malti, 2006, Short Term Opportunities and Challenges for CCS in fossil fuel sector: Current Status of CCS in India, In-session workshop on carbon dioxide capture and storage, UNFCC 24th Session of SBSTA, 20th May 2006, Bonn, Germany.

[11] Goel Malti, 2008, Carbon Capture and Storage: Indian Perspective, in Carbon Capture and Storage R&D Technologies for Sustainable Energy Future, Eds. Malti Goel, B. Kumar and S. N. Charan, Narosa Publishing House, p3-14.

[12] Goel Malti, 2008, Recent Developments In Technology Management For Reduction of CO_2 Emissions In Metal Industry In India, Carbon Dioxide Reduction Metallurgy, Eds. Neale R. Neelameggham and Ramana G. Reddy, A TMS Publication (The Minerals, metals and materials Society), 71-82.

[13] Goel Malti, S.N. Charan, A. K. Bhandari, 2008, CO_2 Sequestration: Recent Indian Research, IUGS in Indian report of INSA 2004-2008, Eds. A K Singhvi, A Bhattacharya and S. Guha, INSA Platinum Jublee publication, pp 56-60.

[14] Kale D. M., 2007, Research Issues in EOR using CO_2, in Carbon Capture and Storage Technology: R& D Initiatives in India, Eds. Malti Goel and Alok Kumar, Joint Publication of Department of Science and Technology and Ministry of Power, pp 129-136.

[15] Sarma Nittala S., Influence of Carbon Sequestration on biogeochemistry of carbon in the ocean, 2007, Department of Physical and Nuclear Chemistry and Chemical Oceanography,Andhra University, Sept.21-22, 2007.

[16] Metz, B., Davidson O., Coninck H., Loos M., Meyer L. Eds., 2005, IPCC Special Report, Working Group III,

[17] Kerr Tom, IEA Energy Technology Scenarios Series 2008, CO_2 Capture and Storage: A Key Carbon Abatement Option.

[18] Pacala Stephen and R. Socolow, 2004, Stabilization Wedges: Solving the Climate Problem for the Next 50 Years with Current Technologies, Science, 305, 968.

[19] Goel Malti, 2009, Recent approaches in CO_2 fixation research in India and Future Perspective towards zero emission coal based power generation, Current Science, 97, 1625-1633.

[20] Analysis of GHG Emissions for Major Sectors in India: Opportunities and Strategies for Mitigation, 2009, Center for Clean Air Policy and IRADe, ICF International, October.

[21] Awareness and Capacity Building in Carbon Capture and Storage, Summary of Lecture Notes, MoES, NESA and INSA Delhi, July 2009.

Chapter 4

Post Combustion Capture of CO_2

Anshu Nanoti and Amar N. Goswami
Indian Institute of Petroleum, Dehradun-248005

SUMMARY

There are three generic methodologies to capture CO_2 from flue gas. These are Post Combustion CO_2 capture, Pre Combustion CO_2 capture and Oxy Fuel Combustion capture. In this chapter background of post combustion CO_2 capture is discussed in detail. Various gas separation technologies including chemical absorption, physical adsorption, membrane separation and cryogenic methods are explained. Merits and demerits of these processes do not allow their field implementation due to significant reduction in the efficiency of a power plant and significant increase in the cost of power generation. Detailed chemistry summarizing the research work being carried out at Indian Institute of Petroleum (IIP), Dehradun on Pressure and Volume Swing Adsorption processes, cost economics and recent breakthrough in CO_2 capture as retrofit options are also described.

INTRODUCTION

World economies are emitting 26 giga tons of CO_2 annually [1]. It has been projected that with increasing population and demand of energy in future cumulative CO_2 amount will touch 9000 giga tons in the atmosphere over this century in the absence of explicit efforts to curtail such emissions. Hence there is need to stabilize the emission of CO_2 in order to avoid severe climate changes which is now an "Inconvenient Truth". As per UN Framework Convention on Climate Change, cumulative CO_2 emissions need not be more than 2600 $GtCO_2$ to 4600 $GtCO_2$. This would mean a substantial reduction and a formidable challenge especially to developing economies.

Industrial CO_2 emissions may come from varied sources like power generation, iron and steel industry, cement plants, petroleum refineries and chemical industries. Sector wise CO_2 emissions from industrial sources are depicted in Figure 1.

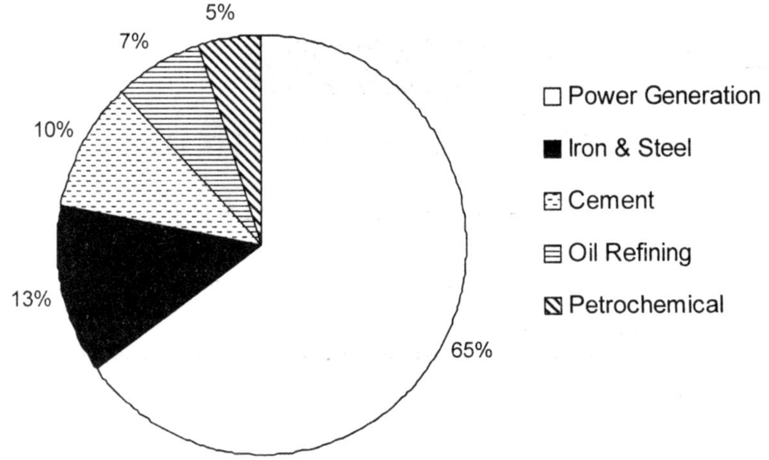

Figure 1: Sources of CO_2 Emissions

Large stationary point sources offer the best opportunities for CO_2 capture and these include fossil fuel based power plants, fuel processing plants etc. World wide there are 8100 large stationary point sources emitting *15 Gt/y CO_2*. The 500 largest emitters among these are coal fired power plants which account for 82 per cent of annual emissions. The past few years have witnessed an increasing interest in retrofitting CO_2 capture technologies in power plants.

CO_2 CAPTURE METHODOLOGIES

There are three generic methodologies [2] that can be used for CO_2 capture from a power plant. These are Post Combustion CO_2 capture, Pre Combustion CO_2 capture and Oxy fuel combustion capture as depicted in Figure 2. Each methodology has its own merits and demerits.

Pre-combustion CO_2 Capture

It is carried out on a process stream and involves fuel decarbonisation through gasification followed by CO_2 removal. During fuel gasification, Syn gas is produced which after water shift reaction, gives a mixture of CO_2 and hydrogen. A CO_2 separation can be implemented to remove the CO_2 and produce high purity hydrogen to fuel the power plant. This method has the advantage of having to treat a stream with high CO_2 concentration and high pressure which are more favourable conditions compared to post combustion capture. The process in principle is same for coal, oil or natural gas feeds, but when coal or oil are used, there are more stages of gas purification. Another problem associated with this method is lower efficiency of gas turbine than conventional turbines.

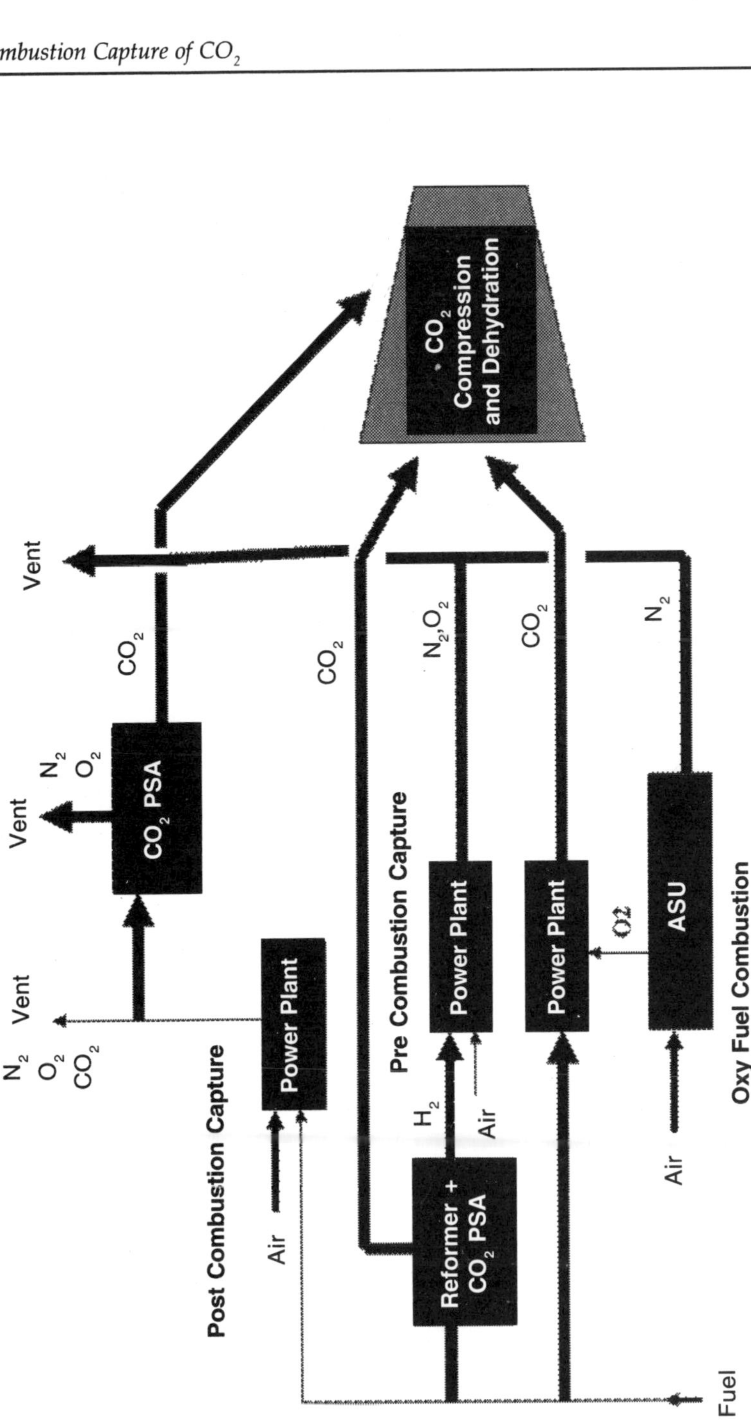

Figure 2: Carbon Dioxide Management in a Power Plant

Oxy Fuel Combustion

Oxy fuel combustion involves replacement of combustion air with oxygen in the furnace. The oxygen is supplied by an Air Separation Unit. Nitrogen, that would be normally present in the flue gas from the furnace is excluded and the flue gas produced is primarily a concentrated stream of CO_2 from which it is recovered by compression. A portion of flue gas however has to be re-circulated back to the furnace to control combustion temperature. The cost of oxygen produced from air is generally high. Concerted efforts are being made to develop low cost oxygen separation methods such as ceramic membranes [3] and/or pressure swing adsorption processes.

Post Combustion Capture

It is an end-of-pipe treatment of the flue gases for removal of the CO_2 present prior to discharge through the stack. A schematic of conventional power plant with option of post combustion capture of CO_2 is depicted in Figure 3. CO_2 levels are generally in the range of 5 per cent to 15 per cent depending on the type of fuel undergoing combustion and the CO_2 must be removed from mixtures with N_2,O_2, moisture and SOx/NOx if present. The flue gas will be at slightly above ambient pressure and at temperatures of around 150ºC.

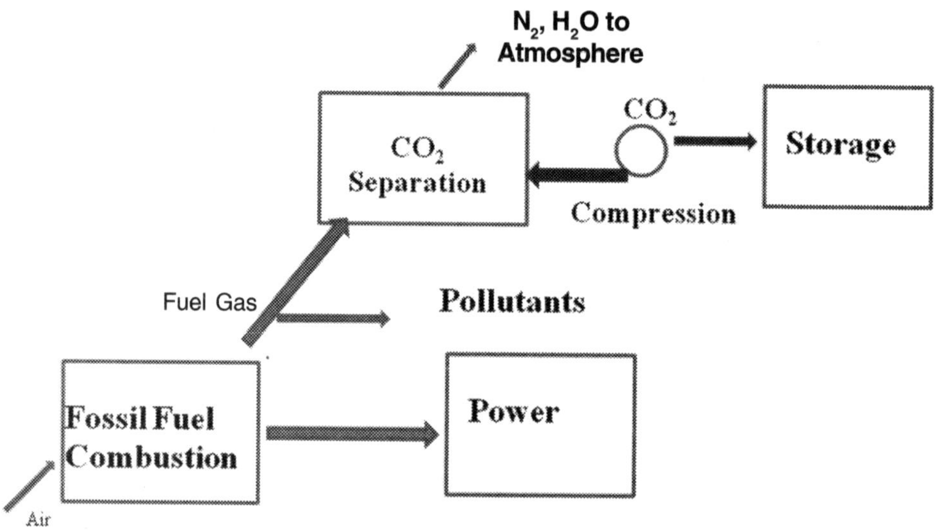

Figure 3: A Conventional Power Plant

A comparison of costs [4] involved in capture of CO_2 from flue gas using above mentioned technologies is given in Table 1.

The cost figures reported in Table 1 appear to favour pre-combustion capture. However it should be kept in view that pre-combustion method is still not widely used in commercial practice and some technical issues still remain unresolved particularly relating to the use of hydrogen as fuel in a power plant. Compared to pre-combustion, post combustion is more mature technology and can be readily adapted in existing power plants without major modification. So there is an urgent need to focus on post combustion CO_2 capture technologies to make them more economical and efficient.

Table 1: Cost Comparison in Three Options

Capture Technology	*Capital Investment*	*Power Output*	*$/kW*
Post Combustion	+23 per cent	–24 per cent	+62 per cent
Oxy fuel Combustion	+14 per cent	–20 per cent	+42 per cent
Pre-Combustion	+7 per cent	–19 per cent	+32 per cent

Source: Howard Herzog/MIT Laboratory for Energy and the Environment, Capacity building programme, May 2007, Pittsburgh, USA

CO_2 CAPTURE PROCESSES

CO_2 capture systems use many of the known technologies for gas separation which are integrated into the three basic systems for CO_2 capture as described above. The technologies are based on absorption, adsorption, membranes and cryogenic separations.

Chemical Absorption

Depending on the partial pressure of CO_2 in the gas mixture, either chemical or physical absorption can be used. Chemical absorption [5] is suited for low to moderate CO_2 concentrations at near atmospheric pressure and is the method of choice of post combustion CO_2 capture. It has been established since last sixty years. Most of the applications of absorption are for natural gas treating and/or for removal of chemically reducing gases. Today it is considered a bridge to CO_2 capture also. The largest reported facility using amine scrubbing is for a throughput 800 t/day with CO_2 capture using two parallel trains. The CO_2 recovery rates are in the range of 85-95 per cent and purity of >90 per cent. The most preferred absorption solvent is mono ethanol amine (MEA). The typical process scheme is given in Figure 4.

In this process CO_2 in flue gas is chemically reacted with an amine in an absorber tower at temperatures of 40 to 60° C. The spent amine is regenerated in a desorber tower by stripping with steam at 100 to 120° C. Typically MEA is used at concentration levels of 30wt per cent in aqueous solutions and CO_2 recoveries are 80-95 per cent and purities of 99.995 per cent. Main concerns in using this technology is corrosion in presence of oxygen in the flue gas and solvent degradation due to presence of SOx, NOx. Large amount of energy is also required for the regeneration of solvent which leads to large parasitic energy losses. Another major concern is large flow rates of flue gas. A typical coal fired power plant of 210 MW capacity emits around 4100 tons per day (tpd) CO_2. Commercial amine processes are handling much lower flow rates [6]. New amine molecules, sterically hindered amines with bulky alkyl groups attached to the amine groups are being tried with advantages for higher capacity and enhanced heat stability. Sterically hindered amines KS-1 and KS-2 are being tried in commercial units by Mitsubishi Heavy Industries [7] for CO_2 recovery and two 450 tpd plants

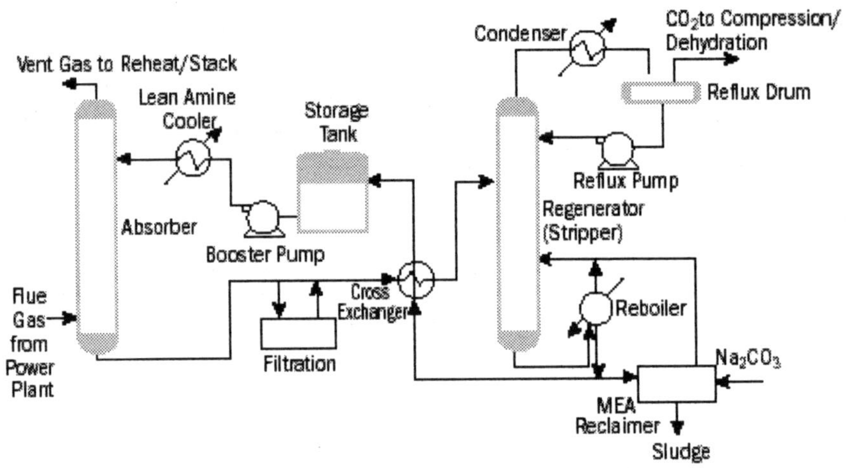

Figure 4: Schematic Diagram of Amine Scrubbing Process

have come up in IFFCO at Aonla (UP). Besides novel solvents, novel designs are also being developed to improve upon existing practices and packing and methods to prevent oxidative degeneration of amine by deoxygenation of solvent solutions.

Adsorptive Processes

Adsorptive processes use adsorbent materials to selectively remove the CO_2 from the gas streams followed by regeneration which can be achieved by reducing the pressure (pressure/Vacuum Swing Adsorption) [8,9] or temperature (Temperature Swing Adsorption) or hybrids. Regeneration by electric swing has also been tried. PSA is a commercially proven technology for hydrogen purification and air separation where the requirement is for producing the weakly adsorbed component in high purity. However for CO_2 recovery for sequestration, high purity of CO_2 which is the strongly adsorbed component is desired and hence PSA with conventional cycles will not work. Thus for this application PSA will require development of new cycles [10]. Again presently available adsorbents will need further development for effective application in CO_2 recovery. Various adsorbents are being studied, these include exchanged zeolites, activated carbon, mesoporous materials, hydrotalcites among others. Typically adsorption capacities on X,Y exchanged zeolites are in the range of 3 to 4.5 mol/kg at 25-30ºC and 1 to 1.2 bar. Figure 5 shows a comparison of reported adsorption capacities with different materials and at different temperatures. The emphasis is on finding adsorbents capable of capturing CO_2 at high temperatures and in presence of moisture. A novel class of mesoporous adsorbent materials show promising results for CO_2 capture at higher temperatures. Typical adsorption capacities in the range of 2.5 to 3.5 mmol/gm have been reported in the literature with amine impregnated mesoporous adsorbents [11,12].

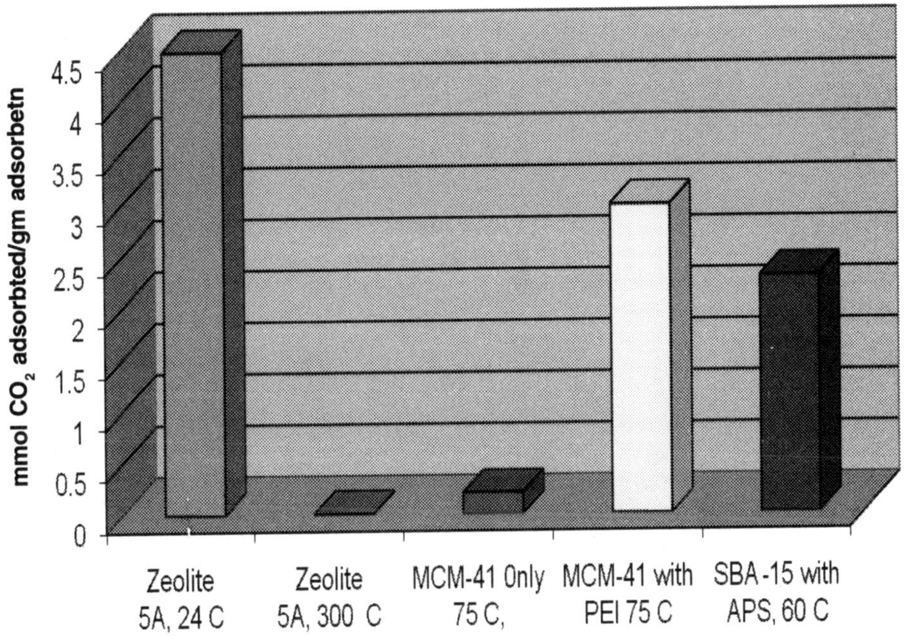

Figure 5: Capacities Reported for CO_2 Capture at 1 to 1.2 Bar

Further development have led to improved enriching cycles which are designed for adsorption of heavy component CO_2. Very significant enrichment of strong adsorptive now becomes possible but there is some penalty in case of lean adsorptive (N_2) purity and hence CO_2 recovery. Figure 6 shows a comparison of enriching reflux cycle with the conventional stripping reflux cycle and dual reflux cycle.

Whereas membrane separations are used commercially for CO_2 removal from natural gas at high pressure and high concentrations, the application to post combustion CO_2 capture is not considered economical due to the low partial pressures of the CO_2 available. The membrane option currently under active development is an integrated system of membrane with absorption, Here a micro porous membrane [13] is used as a support for the absorption solvent in a gas liquid contactor. The CO_2 from the flue gas diffuses through the pores in the membrane into the absorbent, which may be a physical or chemical solvent. Advantages are avoidance of operation problems like foaming, flooding, entrainment, channeling as it occurs in packed beds, large area for gas-liquid mass transfer which reduces equipment size. The desorption can also be carried out in the membrane contactor.

ECONOMICS OF CO_2 CAPTURE TECHNOLOGIES

Cost of CO_2 capture [14] using absorption, adsorption and membranes is given in Table 2.

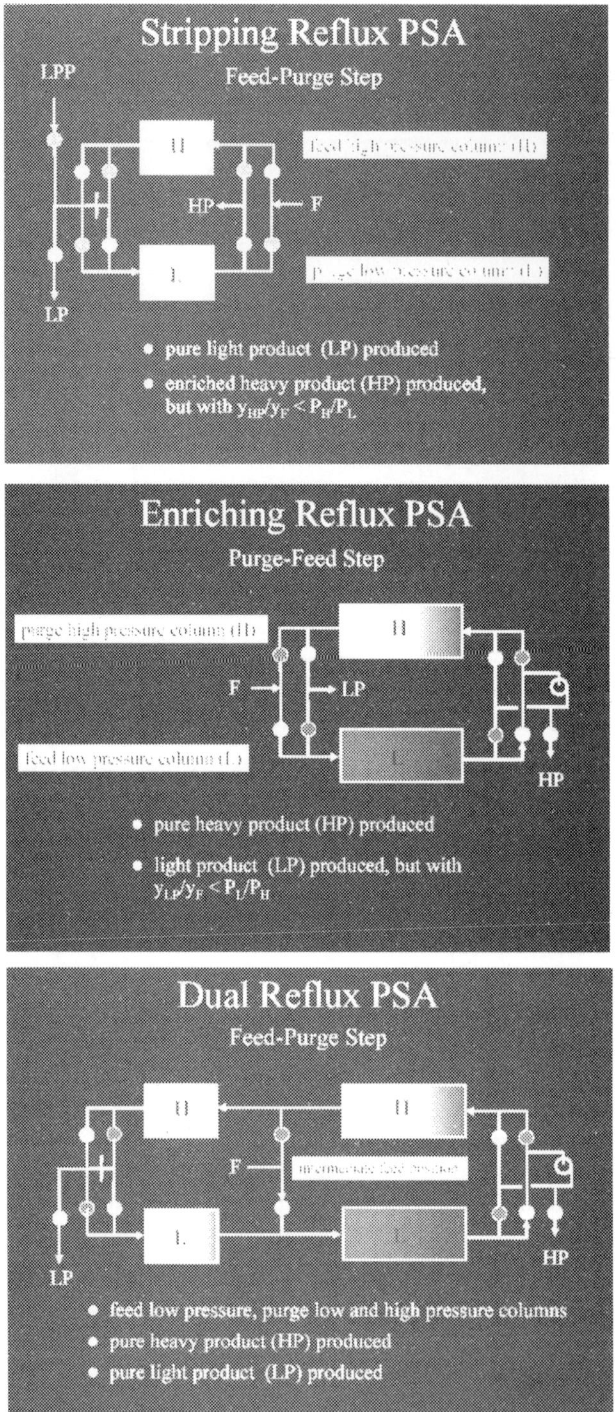

Figure 6: Pressure Swing Adsorption Options

Table 2: Status of Technology and Cost Comparisons in Three Capture Process

Status of Technology	*Chemical Absorption*		*Physical Adsorption*		*Gas Membrane Separation*	
	Baseline Commercial	*State-of-the Art Commercial*	*Baseline*	*Emerging Technology*	*Baseline*	*Emerging Technology*
	MEA Solvent	*KSI Solvent*	*PSA*	*VSA*	*PPO Membrane*	*Co-block Membrane*
CO_2 recovery rate (per cent)	90	90	90	75	90	90
CO_2 purity (per cent)	>98	>98	44	48	43	62
Energy Penalty (per cent)	36	21	47	28	52	45
Capture cost US$/tonne CO_2 avoided	47	34	61	40	78	64

For absorption process, the capture cost is US$ 47/tons of CO_2 avoided. This economics is based on using commercial solvent MEA as the baseline. However the energy penalty is 36 per cent of the total energy output of a 500 MW power plant, out of which 55 per cent is used to regenerate the solvent in the absorption process. The data reported in Table 2 shows a 25 per cent reduction in capture cost from 47 US$ to 34 US$/tons of CO_2 avoided with improved solvent KS-1 of Mitsubishi Heavy Industries (MHI), which still consumes 16 per cent of total power plant output. The capture cost for physical adsorption using Pressure Swing Adsorption as the capture technique is higher than absorption process. These cost figures are based on considering commercial adsorbent and conventional high pressure PSA using Skarstorm cycle.. The product CO_2 obtained is also at low purity, hence additional compression cost is required to compress this diluted CO_2 to 100 bar for sequestration. The cost breakup indicates that around 60 per cent of the cost goes for compression of flue gas to 6 bar and recovered CO_2 up to to 100 bars for sequestering the dilute CO_2 product. It may be pointed out here Skarstrom cycle is suitable for light component (N_2) adsorption. Reduction in capture cost for Vacuum Swing Adsorption (VSA) has been shown with improved CO_2 purity and also compression of feed up to to 6 bar is not required. The VSA cost is comparable to absorption processes. There is still lot of scope in improving the VSA systems with novel adsorbents and with design of new cycles such in a way that are suitable for strong adsorptive component.

CONCLUSIONS

CO_2 can be captured at stationary large point sources such as in the power generation plants and other large energy using industries, which currently account for over half of CO_2 emissions from fossil fuel use. Post combustion capture appears to be the best methodologies which can be easily retrofitted in the existing power plants without major modification. There exists many technologies for CO_2 capture, but integrated operation at a large scale needs to be demonstrated. Based on current technologies CO_2 capture would increase the fuel consumption of power plants by typically about 25 per cent and increase the cost of electricity generation by about 25-50 per cent. There are prospects for reducing the energy consumption and cost of CO_2 capture in future by improvements in technology and wide scale application.

REFERENCES

[1] IEA, 2003, CO_2 emissions from fuel combustion, 1997-2001, IEA/OECD, Paris, France.

[2] CCS Technical Lesson Learned September 2004, www.co2net.com.

[3] Armstrong, P.A., D.L.Bannett, E.P.Foster, and V.E.Stein, Ceramic membrane development for oxygen supply to gasification technologies council, Arlington, Virginia, USA.

[4] Howard Herzog, 2007, MIT Laboratory for Energy and the Environment, Capacity building programme, May 2007, Pittsburgh, USA.

[5] Kohl, A.L. and R.B. Nielsen, 1997, Gas purification 5th ed. Gulf Publishing Company, Houston, Texas.

[6] Arnold, D.S., D.A. Barrett and R.H. Isom, 1982, CO_2 can be produced from flue gas, Oil and Gas Journal, nov., pp. 130-136.

[7] Mimura, T., S.Satsumi, M. Iijima, S.Mitsuoka, 1998, Development of energy saving technology for flue gas carbondioxide recovery by the chemical absorption method and steam system in power plant, in Greenhouse Gas Control Technologies, Proceeding of the 4th International Conference of Greenhouse Gas Control Technologies, Interlaken, Switzerland, Sept., Elsevier Science Ltd.

[8] Ruthven, D.M., S.Farooq, K.S.Knaebel, 1994, Pressure swing adsorption, VCH Publisher Inc., New York.

[9] Penny, Xiao, Jun Zhang, Paul Webley, Gang, Li., R. Singh and M. Marshall, 2008, Capture of CO_2 from flue gas streams with zeolite 13X by vacuum pressure swing adsorption, Adsorption, 14, 575-582.

[10] Diagne, D., M.Goto, T.Hirose, 1995, Parametric studies on CO_2 separation and recovery by dual reflux PSA process consisting of both rectifying and stripping section, Ind. Eng. Chem. Res., 34, 3083.

[11] Hiyoshi, N., K. Yogo, T. Yashima, 2005, Adsorption characteristics of carbon dioxide on organically functionalized SBA-15, Microporous and mesoporous materials, 84, 357.

[12] Xiachun, Xu, S. Chunshan, B.G.Miller and A.W.Scaroni, 2005, Influence of moisture on CO_2 separation from gas mixture by a nanoporous adsorbent based on polyethyleneimine-modified molecular sieve MCM-41, Ind. Eng. Chem. Res., 44, 8113.

[13] Feron, P.H.M. and A.E.Jansen, 2002, CO_2 separation with polyolefin membrane contactors and dedicated absorption liquids: performance and prospects, Separation and purification technology, 27,(3), 231.

[14] Ho, M.T., D.E.Wiley, G.W.Allinson, Reducing the cost of post combustion CO_2 capture, *https://extra.co2crc.com.au/modules/pts2*

Chapter 5

Carbon Dioxide Sequestration Over Mixed Metal Oxide Adsorbents at Higher Temperatures

A.G. *Gaikwad*

CE& PD Division, National Chemical Laboratory, Pune- 411008, India

SUMMARY

The carbon dioxide sequestration from the thermal power stations, burning gas and oil fields are important. The increase in carbon dioxide level beyond threshold value in the atmospheric level causes the adverse effect in the climate change due to the imbalance of gaseous concentrations in the atmosphere. Therefore, there is a need to maintain the appropriate (suitable) concentration of gaseous component in the atmosphere. The carbon dioxide is one of the components in the atmosphere that should be kept within the threshold value. The present exploration is dealing with the selective capture of carbon dioxide with metal oxides in the temperature range 100 to 300°C and 450 to 750°C. The carbon and molecular sieve 13X and 5A adsorbents have low selectivity at ambient conditions. With the increase in the temperature, these adsorbents loose their carbon dioxide capture capacity. Therefore, there is need to explore different metal oxide adsorbents which will selectively sequester carbon dioxide from the other gases at lower and higher temperatures. The CO_2 sequestration was studied at about 1000°C over calcium oxide. The sintering of calcium oxide has been avoided by the addition of sodium at high temperature. The composite materials include combination of sodium, potassium, magnesium, calcium and aluminum oxides. These materials are operated high temperatures 500-800°C. This chapter covers the applications of carbon dioxide, methods of sequestration, different solid adsorbents at 100.to 300°C and 450 to 750°C. It also includes the sequestration of carbon dioxide over silver containing adsorbents, calcium aluminum oxide, and calcium silicate and lithium zirconate.

INTRODUCTION

The sequestration of carbon dioxide from the flue gases having N_2, O_2, CH_4, and H_2 is important in order to avoid the increase in the concentration of carbon dioxide in the atmosphere. Therefore, there is a need to develop the mixed metal oxide adsorbents which will selectively and reactively capture the carbon dioxide from the flue gases. Lithium oxide containing complex materials, such as, lithium silicate and lithium zirconate show the reversible absorption of CO_2 gas at high temperatures. These are the composite materials for the segregation and recovery of CO_2 gas at high temperatures [1]. The sintering of magnesium and calcium oxide is restrained in the presence of aluminum. In addition to this, the adsorption of CO_2 is tremendously increased over magnesium and calcium oxide in the presence of aluminum.

The CO_2 sequestration was studied by the quantitative substitution of the divalent cation and interlayer charge compensating anions in various hydrotalcite-like compounds (HTlcs). The CO_2 adsorption rate and capacity were measured at elevated temperature 330°C over these HTlcs [2]. The adsorption capacities of carbon dioxide have been studied by taking into account the main factors such as aluminum content, anion type, water content, and heat treatment temperature. These factors mainly influence their adsorption capacity at high temperatures. The adsorption of carbon dioxide is favored in presence of carbonate anion in comparison with that of OH^-. The CO_2 adsorption capacity is also improved in the presence of low content of water. In addition to this, the carbon dioxide adsorption capacity is also dependent on the micro-porous volume, interlayer spacing and layer charge density of the hydrotalcite-like compounds [3]. However, the carbon dioxide adsorptions reported over Mg-Al-CO_3 and Ca-Al-CO_3 hydrotalcites at temperature 400°C were 2.29 and 1.79 mmol/g. The carbon dioxide sequestration has been reported over the carbon with or without modification [4-6]. The carbon dioxide capture has been carried out over the modified MCM-41 and Y zeolite with alkali metal [7-10]. The enhanced adsorption of carbon dioxide has been reported over the small amount of alkali or alkaline halogen salts, carbonate, hydroxide, perclorate, bromate or iodate, hydrogen peroxide, phosphate etc supported over the solid adsorbents [11]. The adsorption of carbon dioxide at high temperature has been reviewed over different adsorbents, mixed metal oxides derived from hydrotalcite-like compounds and lithium silicate [12-14]. The stability of hydrotalcite like compounds ceases above 500°C. The adsorption of carbon dioxide over the hydrotalcites is not up to the mark. The investigations of carbon dioxide capture over calcium aluminum oxide, calcium silicate and lithium zirconate at high temperatures 450 to 750°C are lacking. Therefore, this research work reports the carbon dioxide capture over calcium aluminum oxide, calcium silicate and lithium zirconate at different experimental conditions.

METHODOLOGY

Chemicals were used of analytical grade. Silver nitrate, silver oxide, lithium nitrate, calcium nitrates, zirconyl nitrate, aluminum nitrates, ammonium carbonate and ammonium hydroxide used were from Merck India Ltd. The calcium aluminates, calcium silicate and lithium zirconate were prepared with the variation of mole ratio of Ca/Al, Ca/Si/Li/Zr $[M_{1-x}Y_x(OH)_2(CO_3)_{x/2}\cdot yH_2O]$. Where, M = Ca or Li and Y = Al or Si or Zr. The M (1-x mole) and Y (x mole) nitrate solution with required M/Y mole ratio was prepared in 250 ml distilled water (A solution). This mixture solution was stirred to get the complete homogeneous solution. The ammonium carbonate(x/2 mole) and ammonium hydroxide [2.2 for Ca and 1.1 for Li and 3.3 for Al, 4.4 for Si and 2.2 for Zr (mole)] solution was prepared in 500 ml distilled water (B solution). The solution (A) of M and aluminum nitrate or zirconium nitrate or lithium nitrate and ammonium carbonate and ammonium hydroxide (B) were mixed together with slow drop wise addition and simultaneously stirring to get complete precipitation in a glass reactor. In the case of calcium silicate, the tetraethyl ortho-silicate solution was taken in glass reactor. The precipitated solution was agitated for 12 h and then the precipitate was separated by filtration. The solid mass was dried at 100°C for 12 h. Then the dried mass was calcined at 800°C for 3h. The particles of adsorbents were prepared of–22 to–30 meshes.

The low temperature (300 K) sol–gel synthesis of the silver silicate (xAg_2O–[1–x] SiO_2 [x = 0.1–0.8 in steps of 0.1]) was carried out as follows. The chemicals were mixed according to their molar weights ratio for the preparation of different compositions of the silver silicate adsorbents. A solution of TEOS in C_2H_5OH is used in mixing with distilled water. This mixture was stirred for about 3 h until the complete hydrolysis was observed, forming a clear solution. A few drops of 0.1 M HNO_3, as a catalyst, were then added. The silver nitrate solution prepared by dissolving in distilled water is then added to TEOS-alcohol-water solution. These two solutions are mixed until a single, clear solution (sol) was formed. All the prepared sols were cast into separate beakers and then allowed to form gels at 313 K. The clear gels were obtained within 72 h and allowed to dry at 313 K for about three to four weeks. The sol was separated by filtration and calcined. 1 wt per cent of potassium chloride and carbonate were used as catalysts by impregnation on the solid adsorbent and then drying and calcining.

The experimental set up was used for the carbon dioxide adsorption as given in the Figure 1. The adsorption of carbon dioxide was carried out over the adsorbent in a quartz reactor in a temperature controlled split furnace. The adsorbent particle (-22 to–30 meshes) with a suitable amount (0.01 to 0.5 g) was taken in a quartz reactor at the center with quartz wool support. The dimension of quartz reactor was OD 6 mm, ID 4 mm and 850 mm length, and with the modification at the center 210 mm length with 10 mm to 20 mm inner diameter for keeping the adsorbent with quartz wool support. The quartz reactor was connected to Gas Chromatography (GC) through sampling valve with the stainless steel tubing connections. The calcined adsorbent was first flushed with helium gas and then with carbon dioxide to remove other gases. The carbon dioxide capture was carried at a constant temperature and pressure over the adsorbent in the reactor at certain time of exposure. Then the adsorbent was

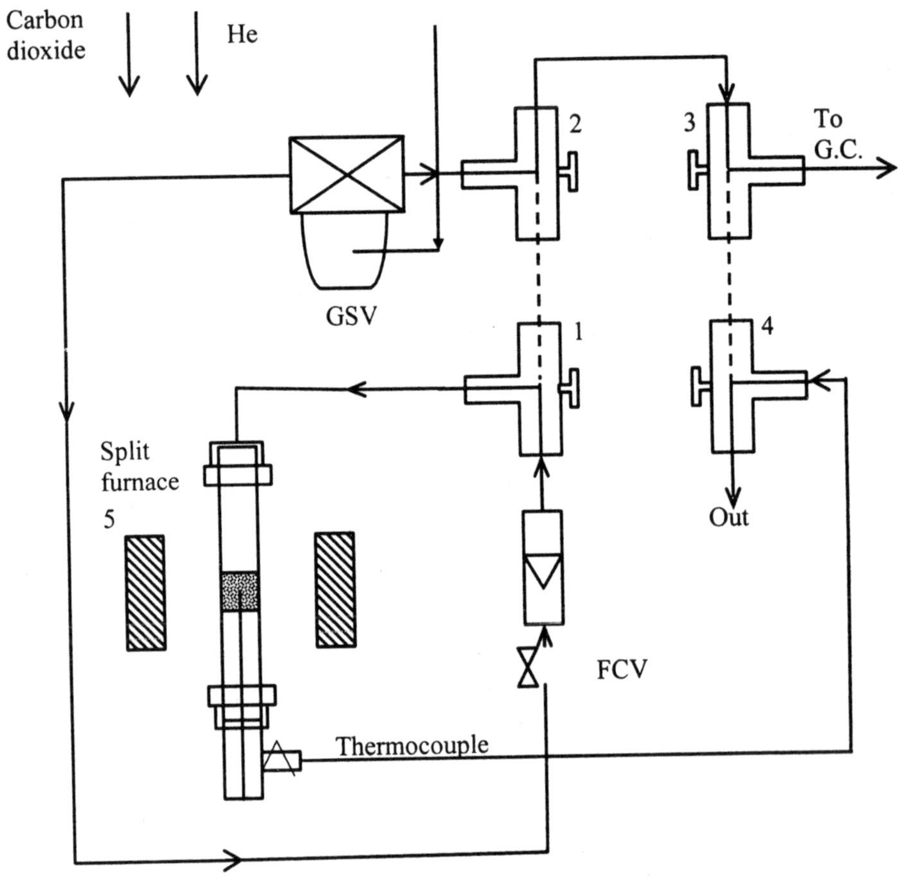

Figure 1: The Schematic Presentation of Carbon Dioxide Sequestration, 1 to 4 Three-way Valves and 5 Split Furnace

flushed with helium gas to remove free carbon dioxide. The adsorbed carbon dioxide was determined by de-sorption with the increase in the temperature. The released carbon dioxide gas was analyzed using the thermal conductivity detector and Porapak-Q column by GC on line analysis. The captured carbon dioxide gas was expressed in terms of mmol/g of adsorbent. The adsorbents were characterized by the measurement of surface area and the XRD pattern.

RESULTS AND DISCUSSIONS

Different Adsorbents and Systems for the Carbon Dioxide Sequestration

The sequestration of carbon dioxide with different systems is illustrated in the Table 1. The carbon dioxide sequestration from the flue gases has been carried out by

Table 1: The Different Adsorbents Used for the Carbon Dioxide Sequestration

Liquid/Solid System	*Extractant/ Adsorbents*	*Scrubber/Removal System*	*Concentration mmol/g*	*Temperature Zone ºC*	*Remarks*
Liquid system	Monoethanol amine	Alkali/NaOH	High	RT	Bulky mass formation
Ionic Liquids	Quaternary ammonium liquid	Alkali/NaOH	High	Up to 250ºC	Bulky mass formation
Grafted/ Supported	Amine grafted on porous solid material/$NaHCO_3$	Temperature desorption	Depend on loading	100 to 250ºC	Loss of grafted material
Solid systems	Carbon	Temperature desorption	2.5 mmol/g 1.74 mmol/g 1.29 mmol/g 0.89 mmol/g	25ºC 1 atm CO_2 partial pressure 30ºC 56ºC 75ºC	Low selectivity for CO_2 in presence of other gases
	Molecular sieve 13 X, 5 X	Temperature desorption	3.69 mmol/g Negligible	25ºC, 1 atm CO_2 partial pressure 200ºC	Low selectivity for CO_2 in presence of other gases
	Supported membrane $NaHCO_3/Na_2CO_3$	Temperature desorption	Depend on loading	450- 600ºC	System for membrane mechanism
	CaO/MgO	Temperature desorption	0.2 mmol/g CaO	400ºC Need high temperature 1000ºC	Sintering
	Li_4SiO_4, Mg_2SiO_4 Ca_2SiO_4 Ag_4SiO_4	Temperature desorption	5 to 30 mmol/g	400–800ºC 150–250ºC	High energy

Contd...

Table 1–Contd...

Liquid/Solid System	*Extractant/ Adsorbents*	*Scrubber/Removal System*	*Concentration mmol/g*	*Temperature Zone ºC*	*Remarks*
	$LiAlO_2$ $MgO–Al_2O_3$, $CaO–Al_2O_3$,	Temperature desorption	5 to 15 mmol/g	400–800ºC	High energy
	Li_2ZrO_3 $MgO–ZrO_2$ $CaO–ZrO_2$	Temperature desorption	5 to 10 mml/g	400–800ºC	High energy
	$ZnO–Al_2O_3$, $CuO–Al_2O_3$, $MnO–Al_2O_3$, $NiO–Al_2O_3$, $CoO–Al_2O_3$, $Ag_2O–Al_2O_3$	Temperature desorption	5 to 15 mmol/g	150–350ºC	Low energy

using liquid and porous solid systems. Amines are used in the liquid membrane or solvent extraction systems for the carbon dioxide sequestration. However, ionic liquids are also most promising for sequestration of carbon dioxide. Grafted or supported amines over porous solid adsorbents have been used for the capture of carbon dioxide. Carbon and molecular sieve 13 X or 5 A are being used for the sequestration of carbon dioxide at room temperature. However, their capturing capacity decreases with the increase in temperature from room temperature to 200°C. Moreover, they are not selective adsorbents for the carbon dioxide sequestration in presence of other gases. Magnesium and calcium oxides have low adsorption capacity for the sequestration of carbon dioxide and in addition to this their sintering restrains the increasing adsorption capacity. The chemisorption of carbon dioxide over the mixed metal oxides can be used for the sequestration of carbon dioxide from the flue gases. The different mixed metal oxide systems are illustrated in the Table 2. Among them, combination

Table 2: The Metal Silicates, Metal Aluminum Oxides, Metal Zirconate and Metal Oxides Explored for the Carbon Dioxide Adsorption

Sl.No	*Mixed Oxides*	*Reactions*	*CO_2 mol*	*Decom.ºC M_xCO_3*
		Metal silicates		
1.	Li_4SiO_4	$Li_4SiO_4 + 2CO_2 \leftrightarrow 2Li_2CO_3 + SiO_2$	2	723
	Li_2SiO_3	$Li_2SiO_3 + CO_2 \leftrightarrow Li_2CO_3 + SiO_2$	1	723
2.	Mg_2SiO_4	$Mg_2SiO_4 + 2CO_2 \leftrightarrow 2MgCO_3 + SiO_2$	2	350
	$MgSiO_3$	$MgSiO_3 + CO_2 \leftrightarrow MgCO_3 + SiO_2$	1	350
3.	Ca_2SiO_4	$Ca_2SiO_4 + 2CO_2 \leftrightarrow 2CaCO_3 + SiO_2$	2	825
	$CaSiO_3$	$CaSiO_3 + CO_2 \leftrightarrow CaCO_3 + SiO_2$	1	825
4.	Ag_4SiO_4	$Ag_4SiO_4 + 2CO_2 \leftrightarrow 2Ag_2CO_3 + SiO_2$	2	210
	Ag_2SiO_3	$Ag_2SiO_3 + CO_2 \leftrightarrow Ag_2CO_3 + SiO_2$	1	210
		Metal aluminates		
5.	$LiAlO_2$	$2\ LiAlO_2 + CO_2 \leftrightarrow Li_2CO_3 + Al_2O_3$	1	723
6.	$Mg_2Al_2O_5$	$Mg_2Al_2O_5 + 2\ CO_2 \leftrightarrow 2\ MgCO_3 + Al_2O_3$	2	350
7.	$Ca_2Al_2O_5$	$Ca_2Al_2O_5 + 2\ CO_2 \leftrightarrow 2\ CaCO_3 + Al_2O_3$	2	825
		Metal zirconates		
8.	Li_2ZrO_3	$Li_2ZrO_3 + CO_2 \leftrightarrow Li_2CO_3 + ZrO_2$	1	723
9.	$MgZrO_3$	$MgZrO_3 + CO_2 \leftrightarrow MgCO_3 + ZrO_2$	1	350
10.	$CaZrO_3$	$CaZrO_3 + CO_2 \leftrightarrow CaCO_3 + ZrO_2$	1	825
		Metal oxides		
11.	NiO	$NiO + CO_2 \leftrightarrow NiCO_3$	1	56-57
12.	ZnO	$ZnO + CO_2 \leftrightarrow ZnCO_3$	1	150-200
13.	CuO	$CuO + CO_2 \leftrightarrow CuCO_3$	1	200
14.	MnO	$MnO + CO_2 \leftrightarrow MnCO_3$	1	200
15.	CoO	$CoO + CO_2 \leftrightarrow CoCO_3$	1	335
16.	Ag_2O	$Ag_2O + CO_2 \leftrightarrow Ag_2CO_3$	1	210

of some transition metal oxides (ZnO, CuO, MnO, Ag_2O) with aluminum oxide in the temperature range from 150 to 350°C can be used. Lithium, magnesium, calcium, aluminum mixed metal oxides, and lithium, magnesium, calcium silicates and lithium zirconate could be used at higher temperatures in the range from 450 to 750°C.

Characterization of Adsorbents

The Figure 2 shows the XRD pattern of the calcium aluminium metal oxide (Ca/Al = 3 mol ratio) and lithium zirconate (Li/Zr = 2 mol ratio). The observed XRD pattern shows the different phases of the calcium aluminum oxide and lithium zirconate adsorbents. Table 3 shows the surface area of calcium aluminum oxide and lithium zirconate adsorbents at different mole ratios.

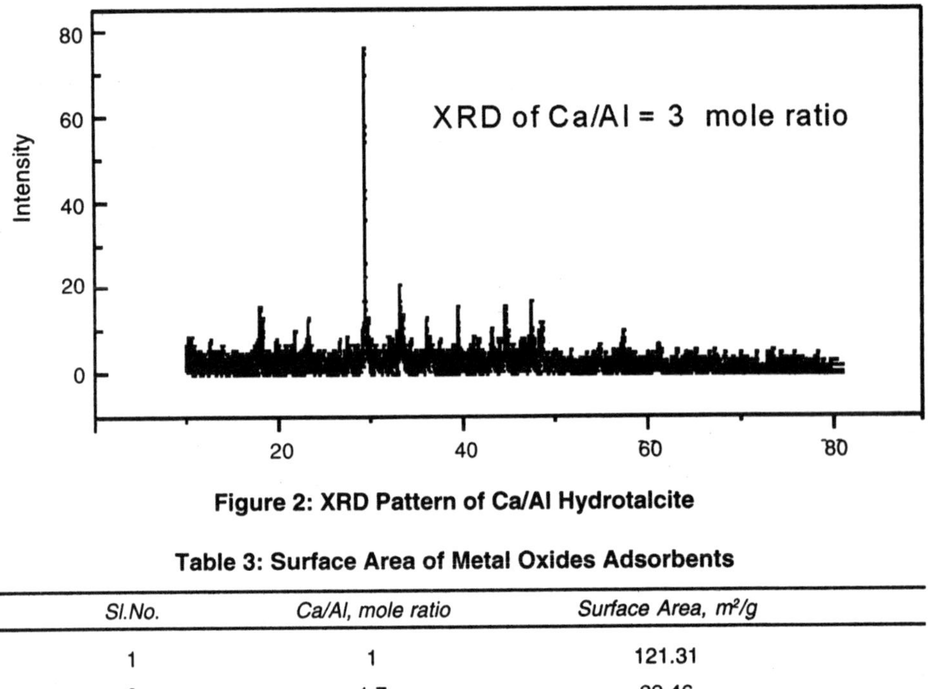

Figure 2: XRD Pattern of Ca/Al Hydrotalcite

Table 3: Surface Area of Metal Oxides Adsorbents

Sl.No.	*Ca/Al, mole ratio*	*Surface Area, m²/g*
1	1	121.31
2	1.7	29.46
3	3	19.11
4	5	8.00

Sequestration of CO_2 Over Silver Containing Adsorbents

The Figures 4–7 show the sequestration of carbon dioxide over the silver oxide and silver silicate adsorbents in the temperature range from 75 to 250°C. The Figure 4 shows the time required for the sequestration of carbon dioxide over the silver oxide and silver silicates. The optimum sequestration of carbon dioxide over silver oxide and silver silicates is illustrated in the Figure 5, and is observed around 160°C. The sequestration of carbon dioxide over silver and silver silicates dependent on the carbon dioxide pressure is shown in the Figure 6. It is higher over silver silicate than

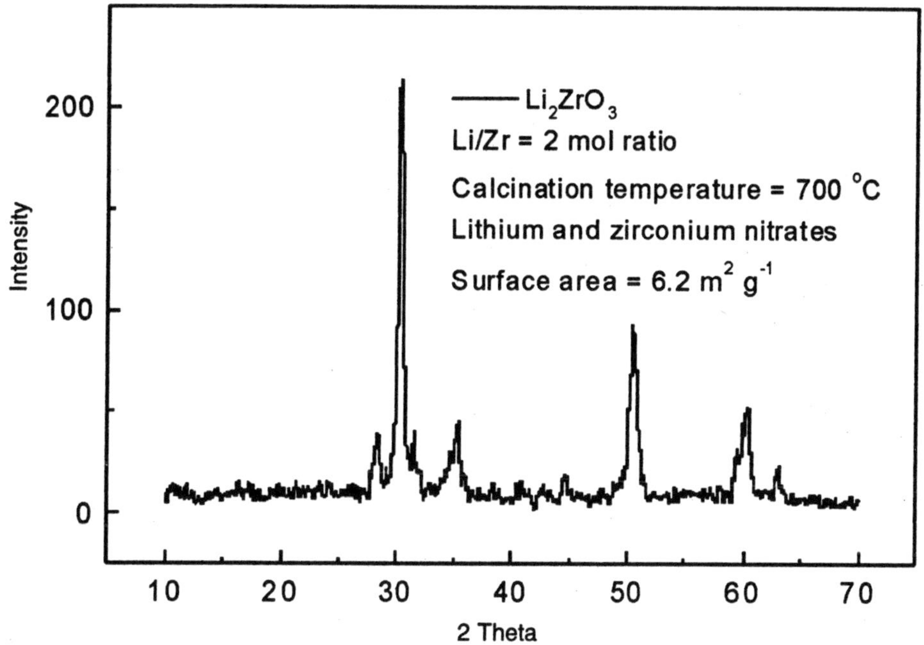

Figure 3: XRD Pattern of Metal Oxides Adsorbents

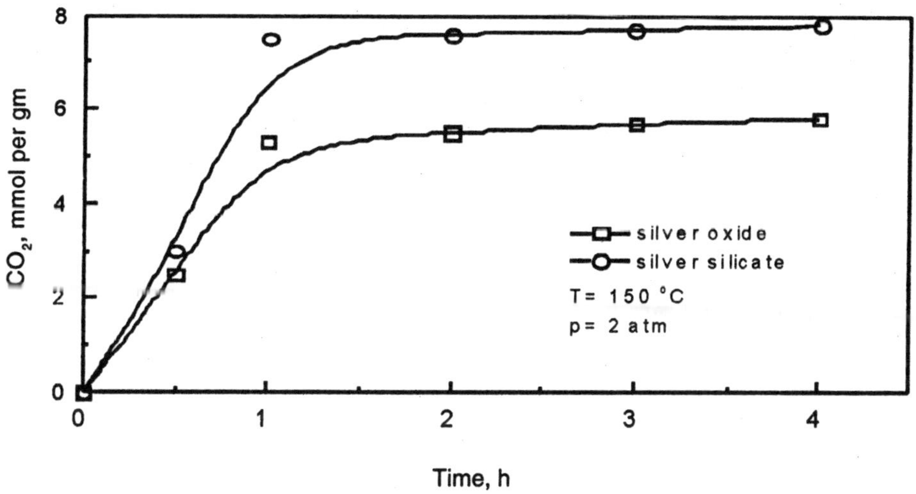

Figure 4: The Effect of Exposure Time of Carbon Dioxide Over Adsorbent

that of silver oxide. The Si to Ag mole ratio of silver silicate has an effect on carbon dioxide capture over silver silicate (Figure 7). The optimum sequestration of carbon dioxide over silver silicate is observed at Si/Ag = 0.3

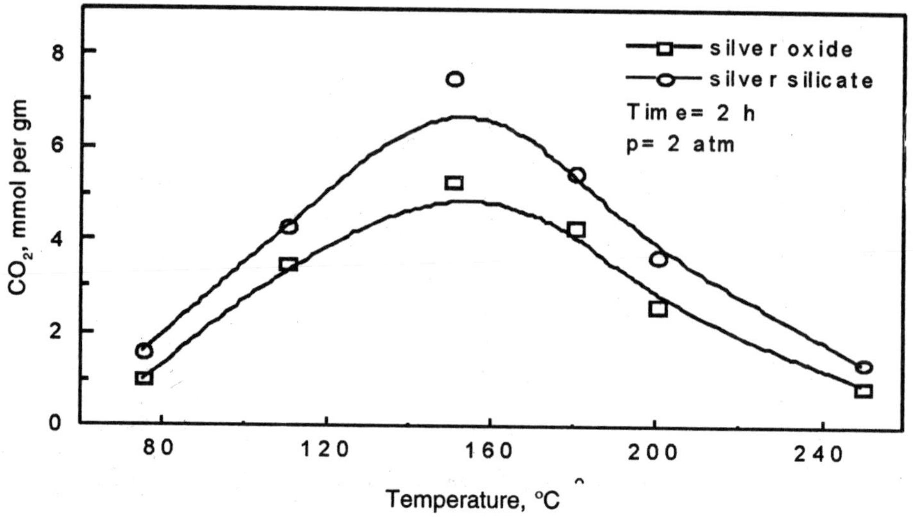

Figure 5: Plot of CO_2 Adsorption vs Different Temperatures

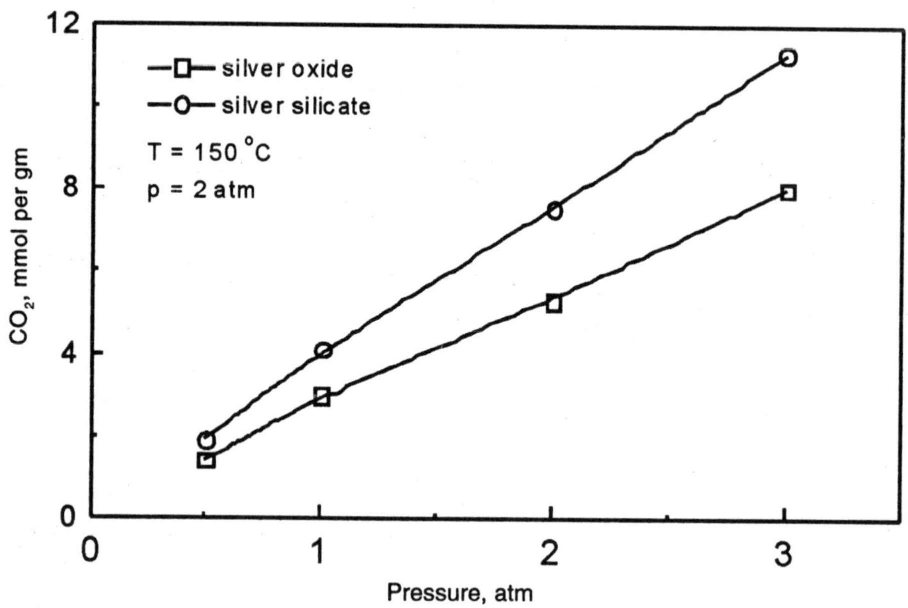

Figure 6: Plot of CO_2 Adsorption vs CO_2 Pressure

Sequestration of CO_2 at Different Temperatures Over Calcium Aluminum Oxide, Calcium Silicate and Lithium Zirconate

The Figure 8 and 9 show the carbon dioxide sequestration over the calcium-aluminum oxides, calcium silicate and lithium zirconate at different temperatures from 40 to 750°C. The carbon dioxide capture was observed in the two temperature

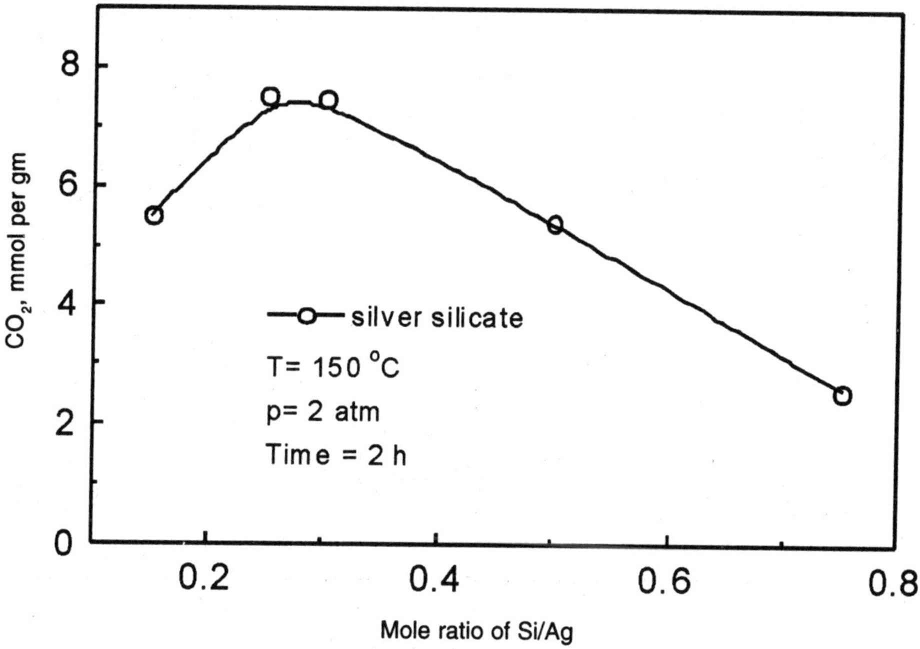

Figure 7: Plot of CO_2 Adsorption vs Mole Ratio of Si/Ag

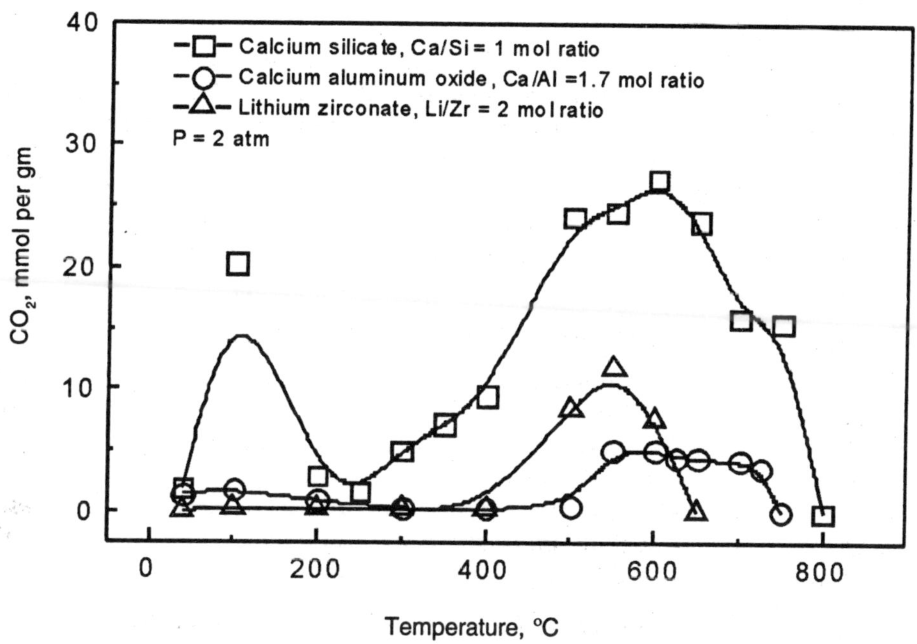

Figure 8: Sequestration of Carbon Dioxide with Respect to Temperature

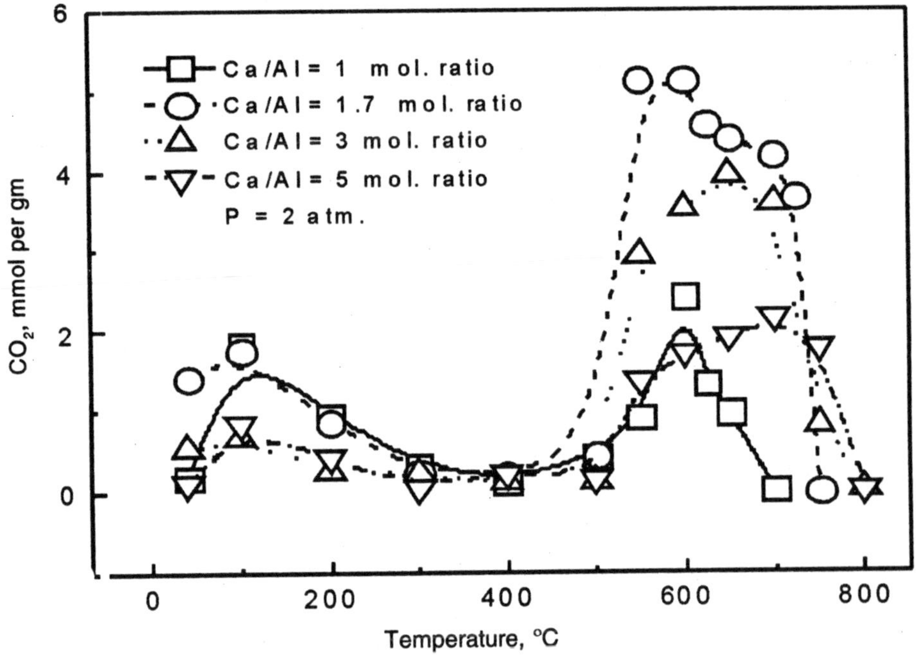

Figure 9: Carbon Dioxide Sequestration Over Ca-Al Oxides at Different Mole Ratio

zones 40 to 400°C and 450 to 750°C. The carbon dioxide sequestration over these adsorbents at higher temperature range is in the range of 5 to 32 mmol/g. The sequestration of carbon dioxide over the calcium silicate was higher than other two adsorbents. The maximum capture of carbon dioxide over the calcium aluminum oxide, calcium silicate and lithium zirconate observed was at 600, 600 and 550° C. After 600°C and 550, the carbon dioxide capture decreases with the increase in the temperature up to 750°C, for the calcium aluminum oxide, calcium silicate and lithium zirconate, respectively.

Sequestration of CO_2 over Calcium Silicate

The carbon dioxide sequestration was explored over calcium silicate at different mole ratio of Ca/Si at different CO_2 pressures at the temperatures 550, 600 and 650°C. Figure 10–12 shows the carbon dioxide capture over the calcium silicate at Ca/Si = 1, 2, 3 mole ratio with the variation of CO_2 pressure from 0.5 to 2.5 atm, respectively. The carbon dioxide sequestration over the calcium silicate increases with the increase in the CO_2 pressure. The calcium silicate shows the maximum capture of carbon dioxide 32.3 mmol/g at 600°C and Ca/Si = 3 mole ratio and p = 2.5 atm.

Sequestration of CO_2 Over Calcium-Aluminum Mixed Metal Oxides

The carbon dioxide was captured over calcium aluminum mixed oxides at different mole ratio of Ca/Al at different CO_2 pressures at the temperatures 550, 600,

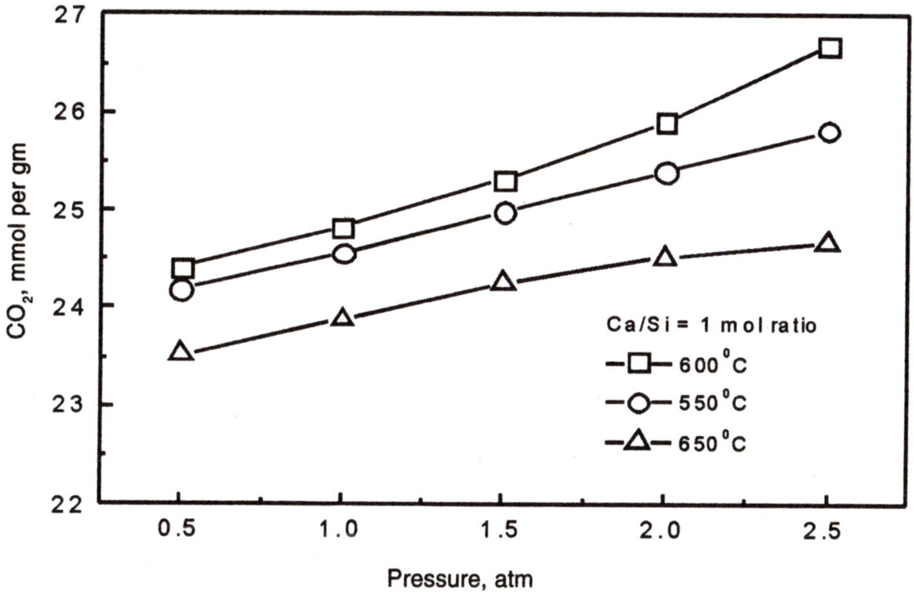

Figure 10: Sequestration of CO_2 Over Calcium Silicate at Different Pressures

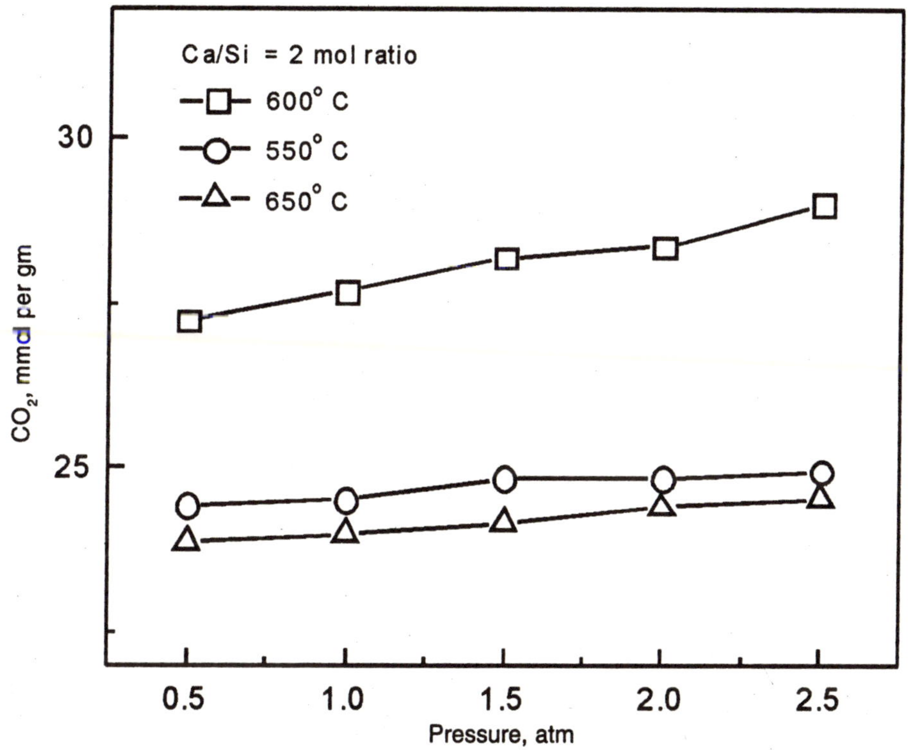

Figure 11: Sequestration of CO_2 Over Calcium Silicate at Different Pressures

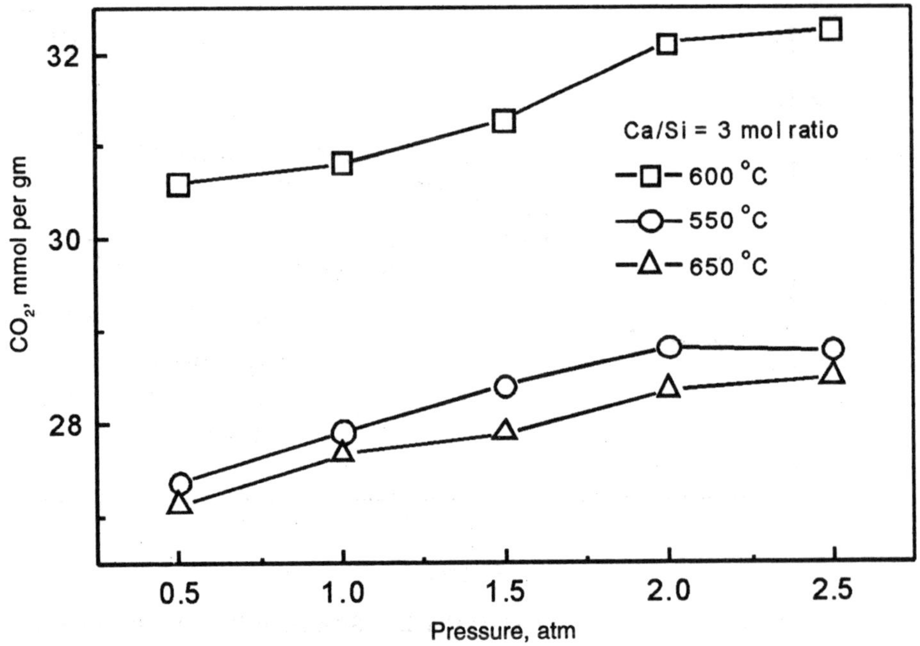

Figure 12: Sequestration of CO_2 Over Calcium Silicate at Different Pressures

650 and 700°C. Figures 13–16 show the carbon dioxide capture over the calcium aluminum mixed oxide at Ca/Si = 1, 1.7, 3 and 5 mole ratio with the variation of CO_2 pressure from 0.5 to 2 atm, respectively. The carbon dioxide sequestration over the calcium aluminum mixed oxides increases with the increase in the CO_2 pressure. The calcium aluminum oxide shows the maximum capture of carbon dioxide 5.2 mmol/g at 600°C and Ca/Al = 1.7 mole ratio and p = 2 atm.

Sequestration of CO_2 Over Lithium Zirconate

The carbon dioxide sequestration was carried out over lithium zirconate at different mole ratios of Li/Zr at different CO_2 pressures at the temperatures 500, 550 and 600°C. Figure 17–19 shows the carbon dioxide sequestration over the lithium zirconate at Li/Zr = 1, 2 and 3 mole ratio with the variation of CO_2 pressure from 0.5 to 2.5 atm, respectively. The carbon dioxide sequestration over the lithium zirconate increases with the increase in the CO_2 pressure. The lithium zirconate shows the maximum sequestration of carbon dioxide 26.5 mmol/g at 550°C and Li/Zr = 3 mole ratio and p = 2.5 atm.

Effect of Metal Ion Mole Ratio Over CO_2 Sequestration

The calcium aluminum oxide, calcium silicate and lithium zirconate were prepared with the variation of their mole ratio from 1 to 5 (Figure 20). Calcium

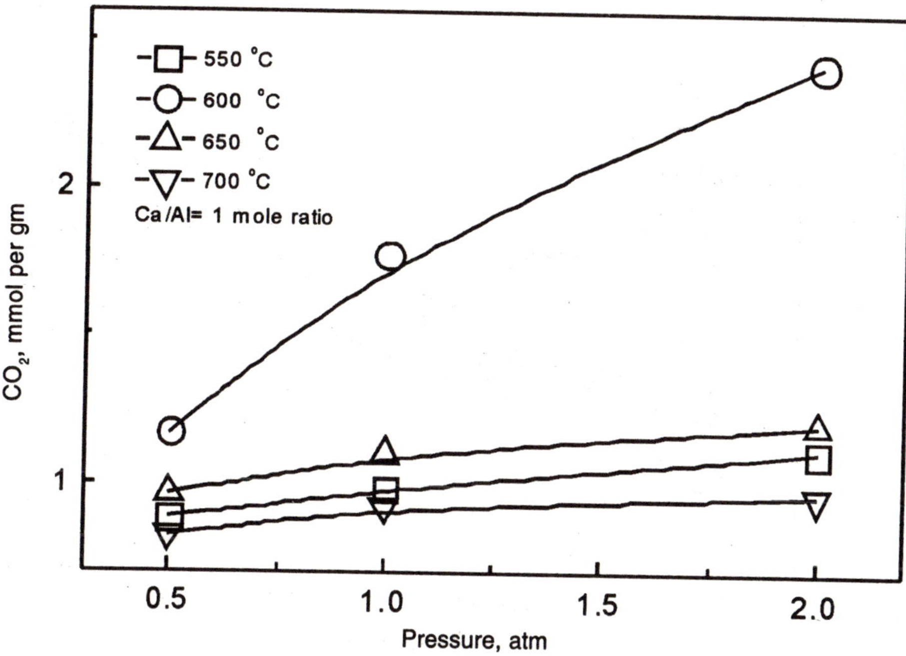

Figure 13: Sequestration of Carbon Dioxide Over Calcium–Aluminum Mixed Metal Oxide Adsorbents (mole ratio 1) at Different Pressures

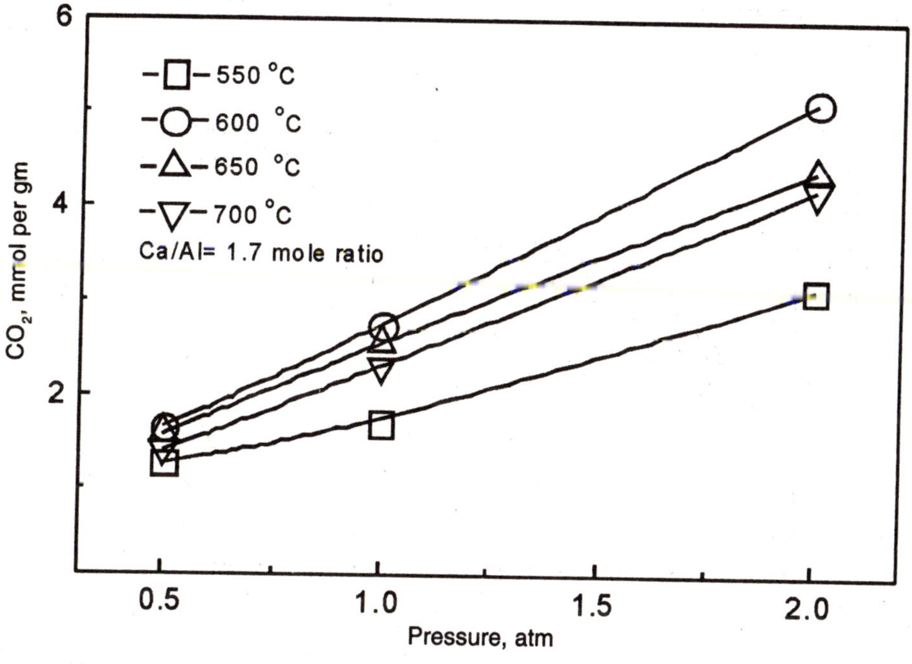

Figure 14: Sequestration of Carbon Dioxide Over Calcium–Aluminum Mixed Metal Oxide Adsorbents (mole ratio 1.7) at Different Pressures

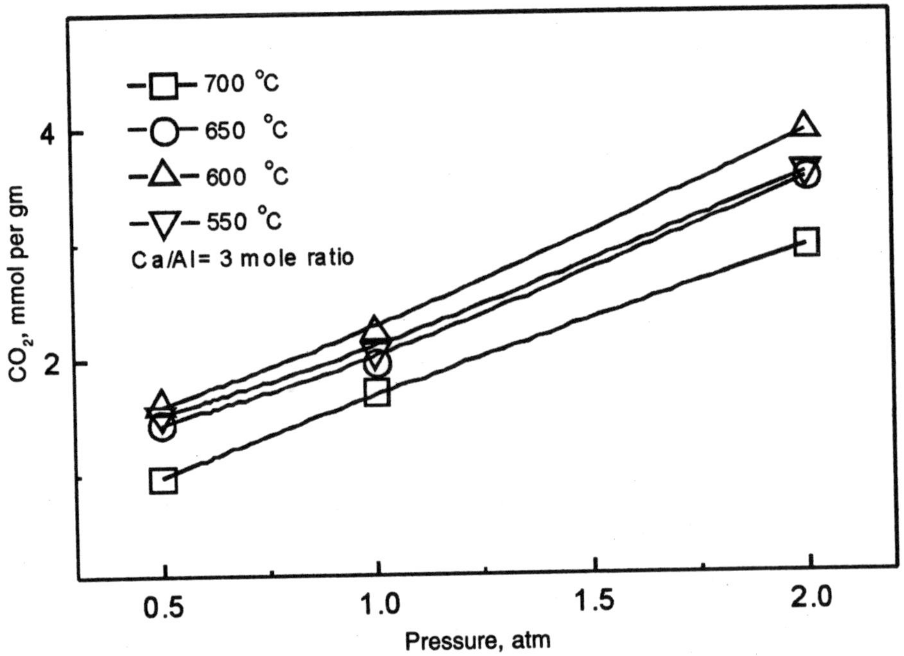

Figure 15: Sequestration of Carbon Dioxide Over Calcium–Aluminum Mixed Metal Oxide Adsorbents (mole ratio 3) at Different Pressures

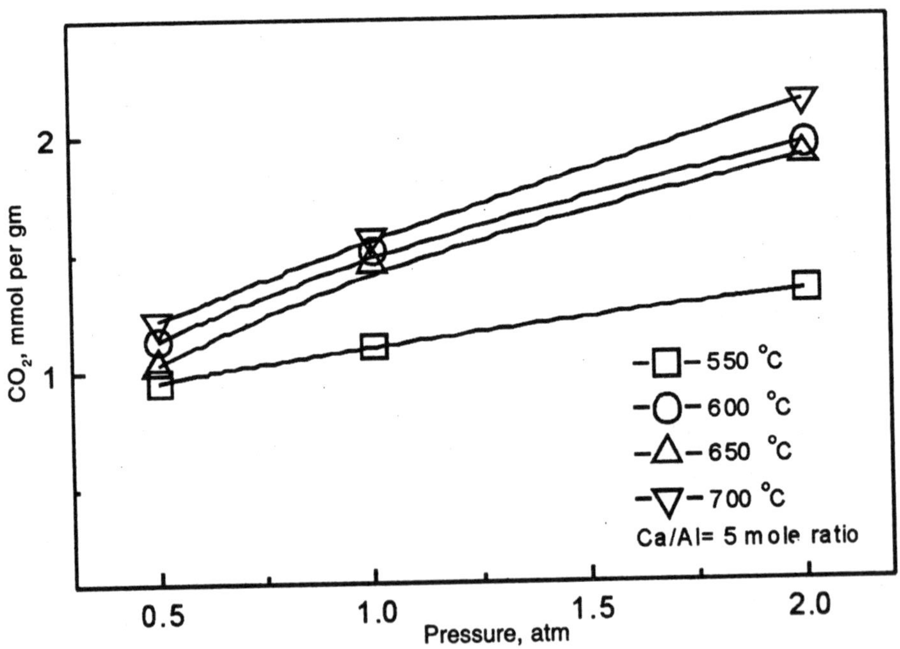

Figure 16: Sequestration of Carbon Dioxide Over Calcium–Aluminum Mixed Metal Oxide Adsorbents (mole ratio 5) at Different Pressures

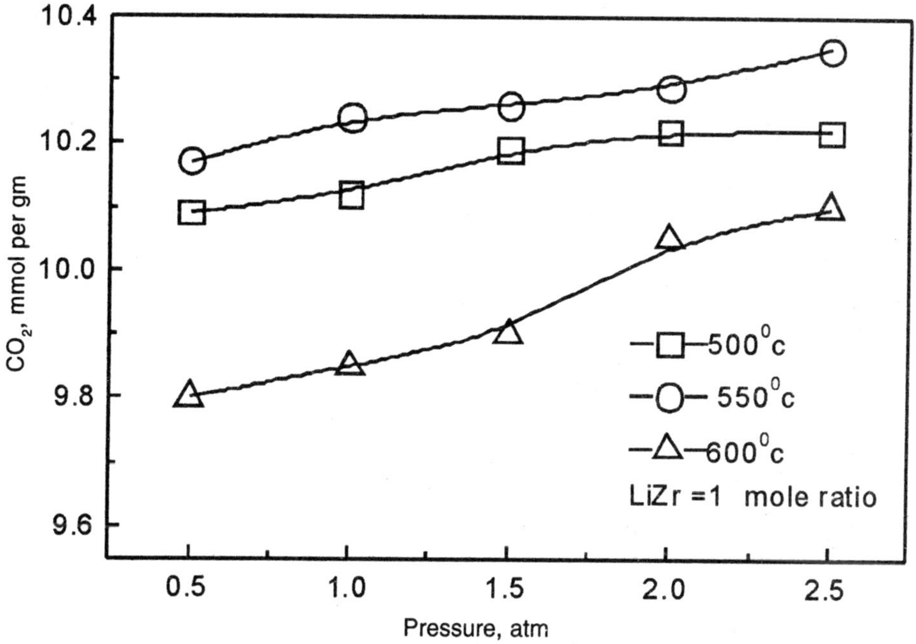

Figure 17: The Carbon Dioxide Sequestration at Different Pressures and Temperatures Over Li-Zr Adsorbent (mole ratio 1)

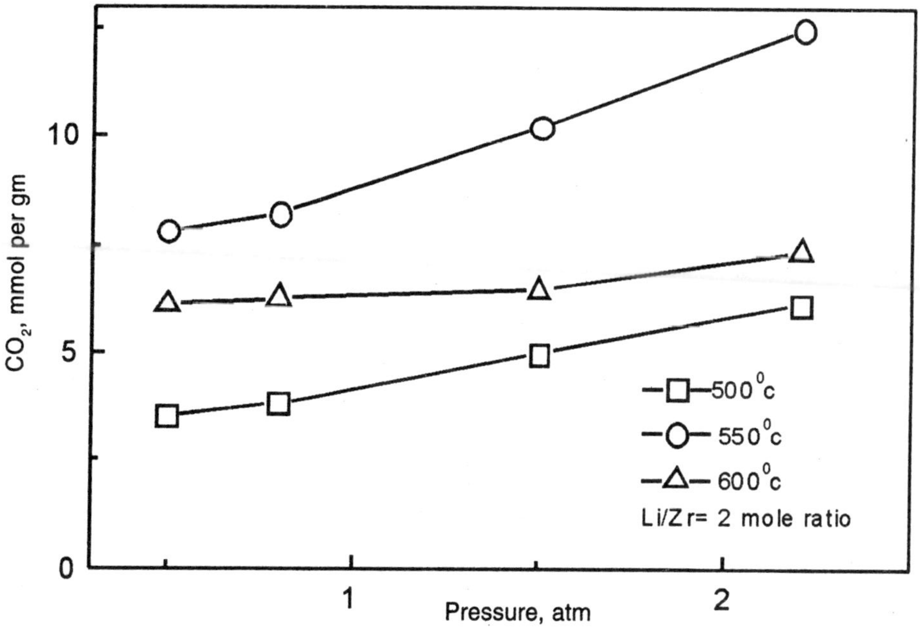

Figure 18: The Carbon Dioxide Sequestration at Different Pressures and Temperatures Over Li-Zr Adsorbent (mole ratio 2)

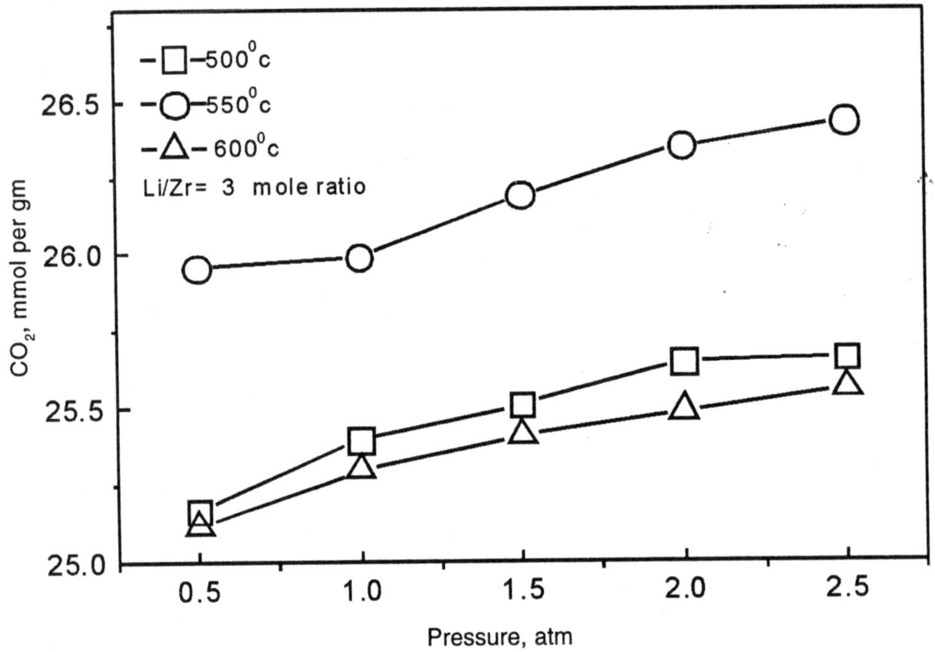

Figure 19: The Carbon Dioxide Sequestration at Different Pressures and Temperatures Over Li-Zr Adsorbent

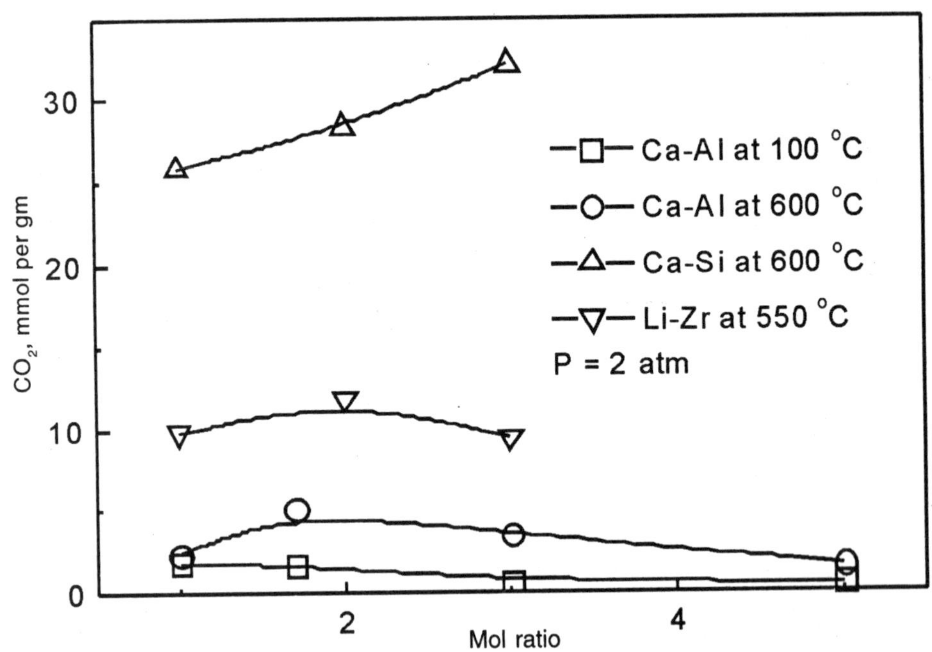

Figure 20: The Effect of Mole Ratio of Ca/Al, Ca/Si and Li/Zr Over the CO_2 Capture

aluminum oxide mole ratio of 1.7 of Ca/Al shows the maximum capture of carbon dioxide. However, the sequestration of carbon dioxide over calcium silicate increases in the studied mole ratio range from 1 to 3. Lithium zirconate shows maximum capture of carbon dioxide at mole ratio 2. Thus the mole ratio of metal oxide in the mixed oxide system has an effect on the carbon dioxide sequestration.

Reusability of Adsorbents

The adsorbents calcium aluminum oxide, calcium silicate and lithium zirconate were tested for their reusability by using same adsorbents for number of times. The

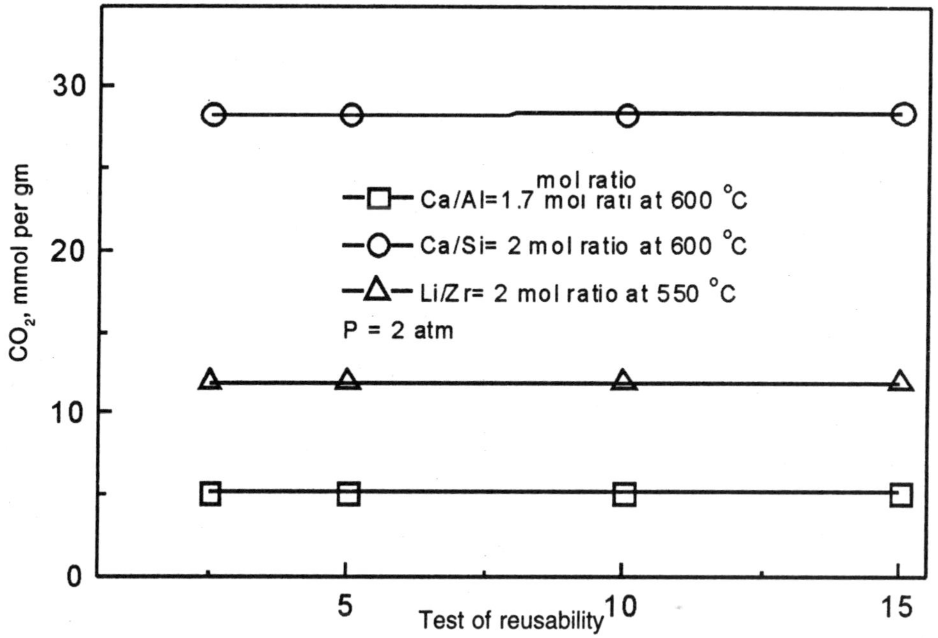

Figure 21: The Test of Reusability of Ca-Al, Ca-Si and Li-Zr Adsorbents for the CO_2 Sequestration

results are shown in the Figure 21. The developed adsorbents show good reusability for15 times uses.

Effect of Promoter

Figure 22 shows the effect of cesium promoter on the capture of carbon dioxide over the calcium aluminum oxide, calcium silicate and lithium zirconate adsorbents. The sequestration of carbon dioxide over these adsorbents enhances in the presence of cesium promoter.

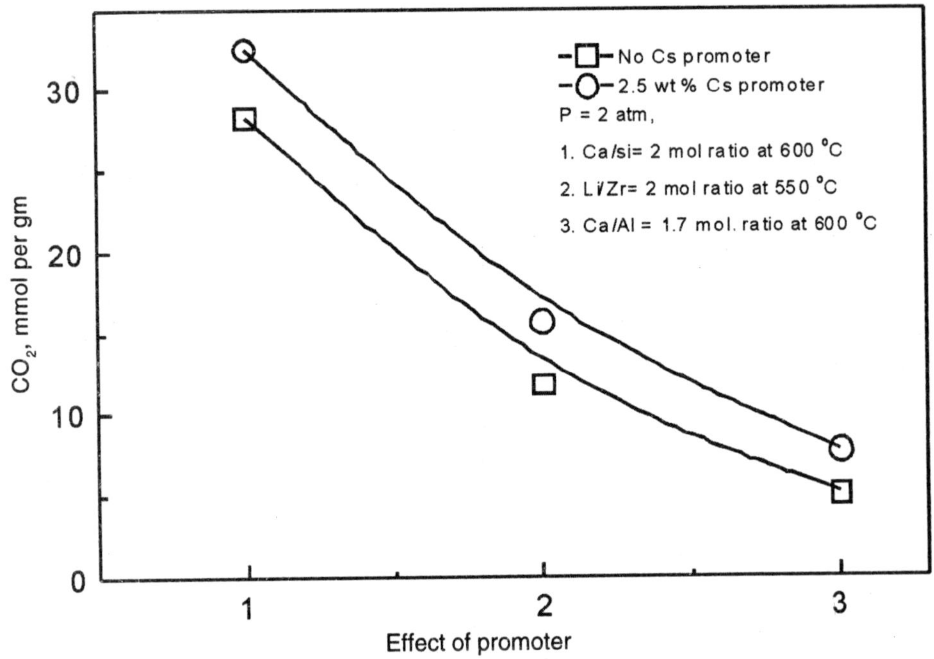

Figure 22: The Effect of Cesium Promoter on the CO_2 Capture Over the Ca-Al, Ca-Si and Li-Zr Adsorbents

CONCLUSIONS

The sequestration of carbon dioxide was studied over the calcium aluminum oxide, calcium silicate and lithium zirconate in the temperature range 40 to 750°C. The sequestration of carbon dioxide over theses adsorbents calcium aluminum oxide, calcium silicate and lithium zirconate at 600, 600 and 550°C was optimum, respectively. The carbon dioxide sequestration at the different mole ratio of metal oxide and at different pressures was explored at their different adsorption temperatures. The reusability of these adsorbents after 15 times uses remains the same.

ACKNOWLEDGEMENT

The author is grateful to the Department of Science and Technology, Government of India for the Research Grant GAP 271526.

REFERENCES

[1] Ueno, S., D. D. Jayaseelan, J. She, N. Kondo, T. Ohji and S. Kanzaki, 2004, Ceramics International. 30, 1031-1034.

[2] Hutson, N.D. and B. C. Attwood, 2008, Adsorption, 14, 781-789.

[3] Yong, Z., V. Mata and A. E. Rodrigues, 2001, Ind. Eng. Chem. Res. 40, 204-209.

[4] Hsin, J. C., D. S. HillWong, C. S. Tan and R. Subramanian, 1997, Ind. Eng. Chem. Res. 36 (7), 2808-2816.

[5] Hayashi, H., J. Taniuchi, N. Furuyashiki, S. Sugiyama, S. Hirano, N. Shigemoto and T. Nonaka, 1998, Ind. Eng. Chem. Res., 37 (1), 185-191.

[6] Na, B., H. Lee, K. Koo, and H. K. Song, 2002, Ind. Eng. Chem. Res., 41(22), 5498–5503.

[7] Xu, X., C. Song, B. G. Miller and A. W. Scaroni, 2005, Ind. Eng. Chem. Res., 44 (21), 8113-8119.

[8] Sircar, S., 1991, Ind. Eng. Chem. Res., 30 (5), 1032-1039.

[9] Husson, S. M. and C. J. King, 1999, Ind. Eng. Chem. Res., 38 (4), 1625-1632.

[10] Daz, E., E. Muoz, A. Vega and S. Ordez, 2008, Ind. Eng. Chem. Res., 47 (2), 412-418.

[11] Johnson, K. S., 1982, Limnol Oceanogr., 27(5), 849-855.

[12] Yong, Z., V. Mata and A. E. Rodrigues, 2002, Sepn. Purif. Technol., 26 (2-3), 195-205.

[13] Wang, X. P., J. J. Yu, J. Cheng, Z. P. Hao andZ. P. Xu, 2008, Environ. Sci. Technol., 42 (2), 614-618.

[14] Sawako, Y., K. Masahiro, K. Kenjie and H. Uemoto, 2001, Nippon Kagakkai Koen Yokoshu., 80, 214-220.

Chapter 6

Oxy Fuel Combustion

M. Soundaraj Raj
AGM, Coal Research, BHEL Trichy

SUMMARY

The chapter covers the definition of oxy fuel combustion and how it differs with conventional method of using air for combustion. The benefits of using oxy fuel combustion over other technologies are brought out. The major countries involved in the development of oxy fuel combustion and the world scenario of the same is provided. The scheme of oxy fuel combustion for a large utility boiler is presented and how it works in a boiler is explained. General features of oxy fuel combustion such as technical barriers, and heat transfer characteristics compared with conventional system. The need for oxy fuel combustion, its prime techniques, burner configuration and typical pay back period for metallurgical application are discussed.

List of test facilities available at BHEL Tiruchirapalli unit and details of each test facility like Solid Fuel Test Burning Test Facility (SFBTF), Fuel Evaluation Test Facility (FETF), Liquid Fuel Burning Test Facility (LFBTF) etc are furnished. The major products Commercialized through In-house R &D are provided.

BHEL's proposal on establishing a pilot plant with objectives and road map is provided. Pre-combustion capture pathways – IGCC power plant with CO_2 scrubbing is furnished.

Oxy fuel combustion

Coal Research
Bharat Heavy Electricals Ltd
Tiruchirapalli

What is Oxy fuel combustion ?

- Oxy-fuel combustion is the process of burning fuel using oxygen instead of air as the primary oxidant, while maintaining the flue gas flow rate unaltered.
- During the process, nitrogen is replaced by CO_2
- Flue gas volume is maintained by recirculation.
- Excess Oxygen in the flue gas is unaltered.
- More than 95 % of flue gas is CO_2

Benefits of Oxy Fuel Combustion

- The complexity of CO_2 capture is drastically reduced
- NOx emission is also drastically reduced
- Reduced cost of CO_2 separation compared to other methods
- Thermal efficiency of the plant also increased.

Grey areas

- Suitability of this technology for TT firing
- Modifications in the burner
- Slagging and fouling for typical Indian coals
- Cost effective Sequestration of CO_2 in India

Major countries involved in the development of Oxy fuel combustion

1. Canada
2. USA
3. Sweden
4. Japan and
5. Australia

Oxy fuel combustion

World Scenario

1. CANMET, Canada, along-with Sandia National Laboratories, California
2. CCSD, Australia
3. Center for Coal Utilisation, Japan
4. Vattenfall University, Sockholm, Sweden

All the above are under pilot scale evaluation stage only.

Oxy-combustion Technology in PC Boilers

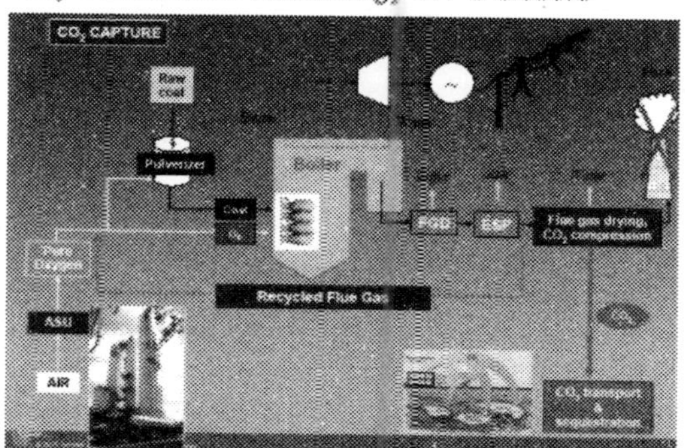

Scheme of Oxy fuel combustion

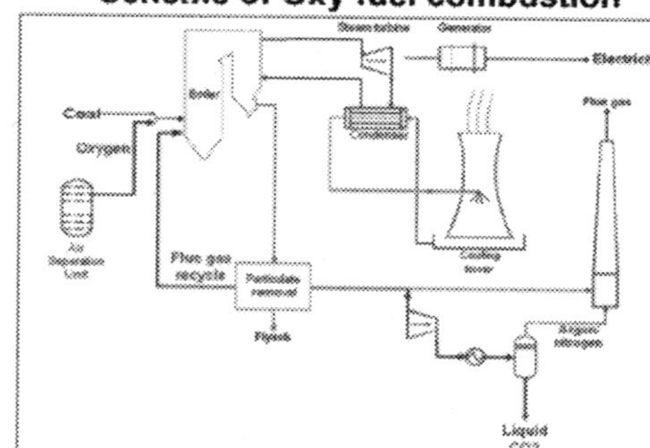

How it works in a boiler ?

- Start the boiler with air and maintain excess oxygen in flue gas around 3.0 %
- Start flue gas recirculation so as to reduce the excess oxygen in flue gas.
- Admit oxygen into the furnace to maintain around 3.0% oxygen in flue gas.
- Increase in steps the quantity of flue gas recirculation and reduce combustion air
- Increase the oxygen quantity to maintain excess oxygen in flue gas
- After sometime the air admission is completely stopped and around 97% of the flue gas will be of carbon di oxide.

Oxy fuel combustion

1. Pilot-scale studies have demonstrated that there are no significant technical barriers to O_2/RFG firing of PF boilers
2. Typically, the optimum O_2 concentration from the ASU for oxy-fuel applications is around 97 - 98%; and the optimum recirculation rate is generally around 70% which yields about 25 – 30% O_2 (vol. %, wet) in the windbox of the boiler, and about 3 - 3.5% O_2 (vol. %, wet) at the furnace exit/AH inlet. At these conditions, flame condition and heat transfer characteristics reasonably approximate those for air-fired PF boilers.

Contd.....

Oxy fuel combustion

3. O_2/RFG combustion yields significant reductions in NOx - typically 25 - 50% lower than for the air-fired case.
4. Preliminary cost evaluations indicate CO_2 capture costs (\$/tCO_2 avoided) and electricity costs (\$/MWh) comparable with other technologies and lower than conventional PF with amine-based post-combustion capture of CO_2.
5. Technical challenges include investigation of flame stability, heat transfer, level of flue gas clean up necessary and acceptable level of nitrogen and other contaminants for CO_2compression, and corrosion due to elevated concentrations of SO_2/SO_3 and H_2O in the flue gas .

Oxy fuel combustion

1. The existing capacity of PF plant worldwide (old and new) is very substantial, and there are plans for a significant number of new PF plants to be installed around the world.
2. The CO_2 capture cost from oxy-fuel is potentially competitive with other emergent technologies.
3. The technical risks associated with oxy-fuel are potentially less than other clean coal technologies because the technology is less complex.
4. In particular countries, the potential for lower capital and operating costs of gas cleaning in oxy-fired PF boilers (deNOx and deSOx) could lead to commercial applications of the technology.

Oxy fuel combustion

Prime techniques

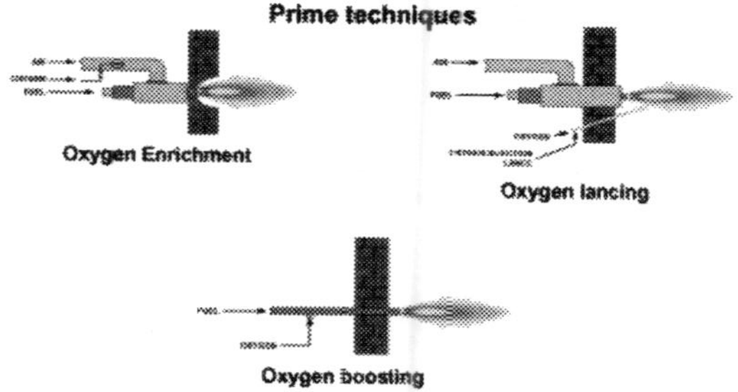

Oxy fuel combustion

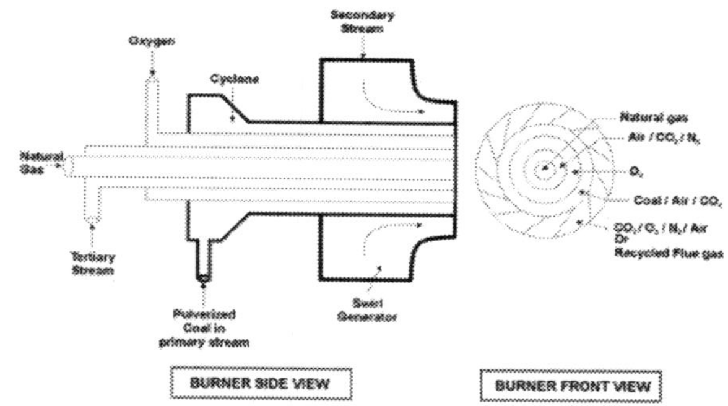

Oxy fuel combustion

Typical payback for metallurgical applications:

Two years for good applications, based on energy savings. Taking other benefits into account will shorten the payback period.

Oxy fuel combustion

Barriers to market Acceptance

1. **Price**
2. **Risk of failure**
3. **Benefits not understood**
4. **Priorities not on benefits of new technology**
5. **Lack of technology awareness**

BHEL'S Initiatives on introducing Oxy fuel combustion in Indian boilers

Bharat Heavy Electricals Ltd, Tiruchirappalli – 620 014

List of Test Facilities available in BHEL Tiruchirapalli.

- Solid Fuel Burning Test Facility
- Fuel Evaluation Test Facility
- Liquid Fuel Burning Test Facility
- Erosion and Abrasion test facilities
- Coal Water Slurry Test Facility
- Pneumatic Conveying Test Facility
- Prototype bubbling fluidised bed combustion boiler – 10t/h
- BHEL-USAID FBC freeboard combustion evaluation facility – 1.6 m.k.cal / h
- Circulating Fluidised bed combustion facility – 4m.k.cal/h
- 6.3 MW CCDP

Coal Research

Bharat Heavy Electricals Ltd, Tiruchirappalli – 620 014

Solid Fuel Burning Test Facility – SFBTF

The SFBTF has a furnace of 10 M kcal/hr with an oil fired furnace with regenerative Air pre- heater to supply hot air to the furnace as well as burner.

Achievements:

1. High Turndown Split Coal Nozzle
2. Low Swirl Scroll Burner
3. Continuous flow coal pump
4. Direct ignition of Pulverized Coal Burner (DIPC)
5. Batch type pneumatic pump
6. Dense Phase Combustion of Powdered Rice Husk
7. Wall burner

Coal Research

Bharat Heavy Electricals Ltd, Tiruchirappalli – 620 014

Fuel Evaluation Test Facility- (FETF)

This is a Fire side Boiler simulator is rated for 756000 kcals/hr (880 KW). The refractory lined furnace has a cross section of 1070 mm x 1230 mm and a height of 7600 mm.

Complex inter-related effects of Fuel characteristic like:

1. Combustion efficiency
2. Heat release pattern
3. Slagging and fouling depositions and their effect on heat absorption and level of emission.. .. are studied

Important Characteristics of:

1. Blended coal firing
2. Co-firing of Pet coke....are studied

Coal Research

Bharat Heavy Electricals Ltd, Tiruchirappalli – 620 014

Liquid Fuel Burning Test Facility - LFBTF

LFBTF has a water cooled shell type furnace of 15 million kcal/hr with auxiliary systems like storage, heating, handling & firing equipments, and an oil fired boiler to supply steam for atomisation.

The achievements are :

- Low NOx low excess air burner
- High Energy Arc (HEA) ignitor
- Air cooled Oil Gun
- Testing of lignite fines-oil-tar slurry
- Testing of Orimulsion
- Testing of Vacuum Residue-New Fuel
- High viscous crude oil

Coal Research

Bharat Heavy Electricals Ltd, *Tiruchirappalli – 620 014*

Coal Water Slurry combustion & handling test facility

Coal Water Slurry preparation handling and combustion test facility to design & develop Coal water slurry preparation and combustion system.

The set up consists of a PC bunker, variable speed rotary feeder slurry preparation and storage tanks, cavity pumps and related instruments.

Achievements

- Established stable CWS preparation with 60-65% solid for high ash Indian coals
- Developed a low cost additive from indigenous sources
- Developed compressed air atomiser of internal mix and external mix types
- Developed a 2.0 Mkcal Swirl burner and flame stability established

Coal Research

Bharat Heavy Electricals Ltd, *Tiruchirappalli – 620 014*

BFBC Development Journey

- Prototype Boiler installed - 1977
- First Commercial Coal fired boiler installed – 1981
- First Rice husk Fired boiler - 1985
- First Washery Rejects Fired boiler – 1987
- First Straw Fired boiler - 1991
- Largest Boiler in operation (165 t/h) - 2001
- First Export order to PT-IBR, Indonesia (3x75t/h)- 2002
- 60 BFBC boilers contracted (55 boilers in Operation)
- Operating experience for more than 2.5 million hours

Coal Research

Bharat Heavy Electricals Ltd, *Tiruchirappalli – 620 014*

CFBC Development Journey

- Test Facility installed – 1991
- Collaboration with LLB (LEE) – 1993
- Technology transfer/Training of Engineers - 1996
- First non –reheat boiler(175 t/h) installed at BILT, Bhigwan - 1998
- First Utility reheat boiler (2x125 MW) installed at SLPP , Mangrol - 1999
- Testing of Lignite – 2001
- 125 MW order received from RVUNL, Giral – 2003
- 75 MW order received from GEB, Kutch – 2004
- 2 x 135 t/h order received from M/s Gujarat Ambuja cement – 2004

Coal Research

Bharat Heavy Electricals Ltd, *Tiruchirappalli – 620 014*

CFBC Development Journey (contd…)

- 2 x 250 MW order received from Neyveli Lignite Corporation – 2005
- 125 MW repeat order received from RVUNL, Giral – 2005
- 2 x 125 MW order received from NLC – Barsingsar firing Rajasthan Lignite– Oct 2005
- 2 x 125 MW repeat order received from SLPP – March 2006
- 1 x 135 t/h repeat order received from M/s Gujarat Ambuja cement - 2006
- 3 x 275 t/h order received from M/s Bharat Oman Refineries Ltd firing 100% Petcoke – Sep 2006
- 18 Boilers Contracted (3 boilers in Operation)

Coal Research

Bharat Heavy Electricals Ltd, Tiruchirappalli – 620 014

Erosion Test Facility:

- Erosion test facility to evaluate the parameters influencing erosion.
- Erosion studies can be made with Velocity up to 50 m/s, dust concentration up to 450 g/m^3 and angle of impact up to 90^0

Achievements :Parameter influencing erosion, wear resistance of materials, evaluation of deferent varieties of hard facing electrodes and wear resistant ceramic material (CERALIN) bends for coal were studied

Abrasion Test Facility:

- Abrasion Test Facility to evaluate abrasiveness of coal and relative abrasion resistance of materials.
- Coal size 16 to +30 mesh can de studied.

Achievements: Relative abrasiveness of various Indian Coals, abrasion resistance of various coatings and a correlation between abrasion index and mill roller wear life were studied.

Coal Research

Bharat Heavy Electricals Ltd, Tiruchirappalli – 620 014

Current Status
pressurised fluidized- bed gasifier

- Pressurized Fluidized Bed Gasification Process Established
- Further work in progress for:
  - Enhancing Gas Conversion Efficiency
  - Reducing Cost

• Coal type tested	High Ash
• Ash content of coal tested	28 – 42%
• Gasification temperature	1050 deg C.
• Gasification pressure	0.67 MPa.
• Cumulative operating hours	4000

Coal Research

Bharat Heavy Electricals Ltd, Tiruchirappalli – 620 014

Major products commercialised through in house development

- ❖Atmospheric Fluidised Bed Combustion boilers
- ❖Circulating Fluidised bed boilers
- ❖Low NOx oil burner
- ❖DIPC system
- ❖BOFA system for NOx reduction

Coal Research

Oxy fuel combustion

Oxy fuel combustion system will be retrofitted into our Fuel Evaluation Test Facility (FETF)

Suitability of Oxy fuel combustion for high ash Indian coals will be studied for flame stability, carbon loss, slagging, fouling and furnace temperature stability etc.

Burner scale up studies will be carried out at our Solid Fuel Burning Test Facility (SFBTF)

Oxy fuel combustion will be demonstrated in a 210/250 MW boiler

BHEL's Road map on Oxy fuel combustion

Rerofitting Oxy fuel in FETF.............................Mar 2010

Test trials of Oxy fuel combustion in FETF...............Dec 2010

Burner scale up studies in SFBTF..........................Mar 2011

Demonstration in a 210/250 MW boiler....................Mar 2013

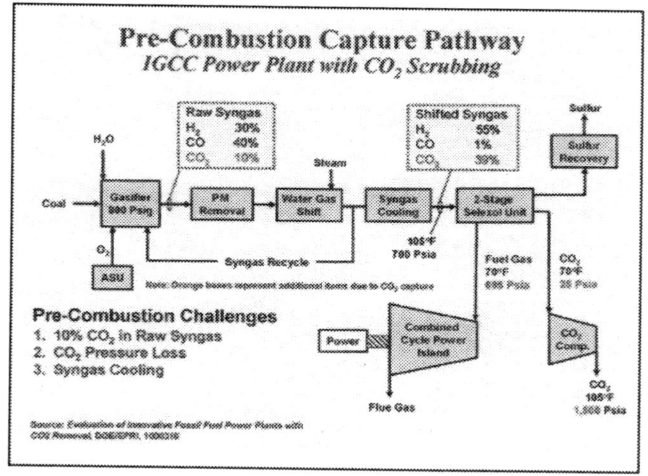

Chapter 7

CO_2 Sequestration and Biofuel Production Using Micro Algal Technology

M. Premalatha*, K.K. Vasumathi, K. Sudhakar

Centre for Energy and Environmental Science and Technology. National Institute of Technology, Tiruchirappalli – 620 015

SUMMARY

Global warming is caused mainly by the emission of carbon dioxide (CO_2), with thermal power plants being responsible for about 7 per cent of global CO_2 emissions.The Use of fossil fuels to meet our energy requirement has resulted in adverse effects on the climate, over dependence on foreign oil and economic uncertainties. The increase in the surface temperature due to global warming, causing catastrophic effects, compels the scientist community to tag it as an issue of prior concern. The most of the environmental group urging for individual, group or community actions against global warming, through switching to alternate energy to hinder the speed of warming up. These methodologies remain virtual. To mitigate these harmful effects, biological alternatives to capture CO_2 are being investigated. The use of carbon sequestration by micro algae is a major tool for reducing atmospheric concentrations of CO_2. This paper highlights the photosynthetic way of fixing CO_2 from flue gas and the various technical and economic challenges involved in it. Current research issues such as design of algae pond and photo bioreactor, algae strain selection, operational problems and harvesting techniques are described.

* Corresponding Author.

INTRODUCTION

Atmosphere is loaded with around 90Mt (million tonnes) of heat trapping substances every day that slowly wrap the earth with an artificial greenhouse gaseous screen. Most anthropogenic carbon dioxide (CO_2) emissions result from the combustion of fossil fuels for energy production. Flue gases from power plant are the main sources of CO_2 emissions which lead to Global warming. To tackle climate change effects CO_2 level should not be allowed to get much higher than 550ppm: the current level is 380 ppm. CO_2 emissions is expected to increase at an annual rate of 3 per cent.Consensus within the scientific and most of the political community is that emission of greenhouse gases is detrimental to the environment and results in worse air quality and alteration of global biological systems. The potential effects of global warming on India vary from the submergence of low-lying islands and coastal lands to the melting of glaciers in the Indian Himalayas, threatening the volumetric flow rate of many of the most important rivers of India and South Asia. In India, such effects are projected to impact millions of lives. As a result of ongoing climate change, the climate of India has become increasingly volatile over the past several decades; this trend is expected to continue. Hence there is a urgent need to solve the problem of CO_2.

THEORETICAL BACKGROUND: PHOTOSYNTHETIC CO_2 SEQUESTRATION AND CAPTURE

The means of preventing the twin catastrophes of energy scarcity and environmental ruin are unclear, but one part of the solution may lie in microbial energy conversion. Limiting the use of traditional fossil fuels in favor of biological sources has been proposed to reduce the amount of harmful greenhouse gases released into the atmosphere because biological sources take up carbon from the environment during photosynthesis, creating a closed carbon cycle.When coal is burned to create electricity, carbon locked in the coal (by plants and animals) over millions of years is released into the atmosphere resulting in a net increase in the total amount of carbon. There is potential to effectively reduce the amount of carbon dioxide and nitrogen oxides released into the atmosphere from many stationary emitters by feeding the carbon-rich flue gas to the algae. The carbon used to create lipids in the algae is still released into the atmosphere upon combustion of the fuel, but the overall amount of carbon has been used twice: once for energy production in a power plant and second to grow algae for transportation fuels. Figure 1 depicts the process of recovery and sequestration of CO_2 using micro algae. As carbon regulations are likely to be set in the future, using algae to consume CO_2 will become even more appealing

Characteristics of Micro Algae

Algae are simple unicellular organisms that produce carbohydrates, proteins and lipids as a result of photosynthesis. Sunlight, water, nutrients and arable land are the major requirements for growing algae. Micro algae have the ability to fix CO_2 using solar energy with efficiency 10 times greater than that of the terrestrial plants with numerous additional technological advantages. Algae are more efficient at utilizing sunlight than terrestrial plants [1], consume harmful pollutants, have

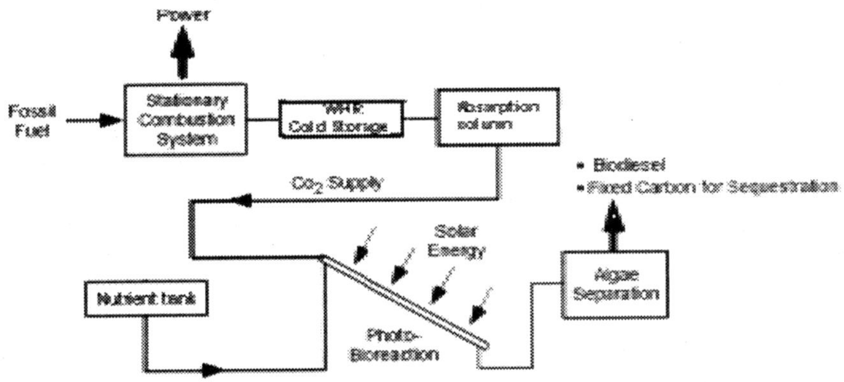

Figure 1: Recovery and Sequestration of CO_2 from Stationary Combustion System by Photosynthesis of Micro Algae

minimal resource requirements and do not compete with food or agriculture for precious resources [2]. India has a unique opportunity for algae production because it contains the basic resources needed to grow algae in abundant quantities: India produces over 170 million metric tons of CO_2 annually; contains abundant saline water; receives abundant sunlight; and has an impressive knowledge base and technical expertise within the energy industry.

Algae have higher growth rates than terrestrial plants, allowing a large quantity of biomass to be produced in a shorter amount of time in a smaller area. Algae growth rates of 10 to 50 g m^{-2} d^{-1} (grams of algal mass per square meter per day) have been published in the literature [3]. Compared to terrestrial plants such as corn and soy, algae have shorter harvest times because they can double their mass every 24 hours.

These short harvest times allow for much more efficient and rapid production of algae compared to corn or soy crops. To illustrate the land requirements for bio fuel, crop production, yields of different oil producing crops can be examined, as shown in Table 1. Compared to terrestrial crops, algae utilize solar energy more efficiently and because they are not limited to one growth cycle per year, they can be harvested much more often.

Table 1: Typical Yields and Land Data [4, 5]

Crop	*Oil Yield (gal/acre-yr)*	*Land Area Needed (million acre)*
Corn	18	2222
Cotton	35	1143
Soyabean	48	833
Canola	127	315
Jatropha	202	198
Oil palm	635	63
Micro algae(15 per cent oil)	1200	33
Micro algae (50 per cent oil)	10000	4

Resource Requirements for Algae Growth

One of the most compelling advantages of using algae as a bio fuel feedstock is that the resource requirements are less intensive compared to other crops and plants. Algae require only a few basic resources to grow successfully: CO_2, water, sunlight and nutrients. Sunlight is normally abundant throughout most of the year and utilized more efficiently than terrestrial crops. CO_2 can be obtained in high concentrations from power plants and industrial processes or at ambient concentrations from the atmosphere. Algae will grow in most water sources with varying pH levels from fresh drinking water, saline or brackish aquifers and wastewater effluent [6]. Brackish, or moderately salty water is abundant and provides a suitable environment and resource for algae to grow in. Algae, by virtue of photosynthesis, are adept at sequestering CO_2 or nitrogen oxides from the atmosphere [7].There is potential to effectively reduce the amount of carbon dioxide and nitrogen oxides released into the atmosphere from many stationary emitters by feeding the carbon-rich flue gas to the algae [8]. Algae are therefore able to fix approximately 1.8 kg of CO_2 for every 1 kg of algae biomass produced [9]. Based on the literature, one can determine that approximately 40 ha of algae ponds are required to fix the carbon emitted from one MW of power generated from a coal plant [10].

Oil Content and Composition of Algae

Algae can be oil-rich organisms. Oil content, the percentage of oil per weight of dry biomass, typically ranges from 20 to 50 per cent depending on the species. This oil is composed of many different types of lipids that can be processed easily into biodiesel, jet fuel or other chemicals. Algae species and their typical oil contents are presented in Table 2 below.

Table 2: Algae Species and Typical Oil Content [9]

Microalga	*Oil Content (per cent dry weight)*
Botryococcus braunii	25-75
Chlorella sp.	28-32
Crpthecodinium cohnii	20
Cylindrotheca sp.	16-37
Dunaliella primolecta	23
Isochrysis sp.	25-33
Monallanthus salina	>20
Nannochloris sp.	20-35
Nannochlropsis sp.	31-68
Neochloris oleoabundans	35-54
Nitzschia sp.	45-47
Phaeodactylum tricornutum	20-30
Schiochytrium sp.	50-77
Tetraseknus sueica	15-23

Compared to terrestrial crops such as corn, soy or even palm plants, algae are far more oil-rich and offer a higher yield of oil per unit of land in a year. Of particular interest are the lipids, which can be processed into valuable fuel products. Lipids produced from algae contain saturated and polar lipids, which are suitable for use as a fuel feedstock and are contained in higher concentrations than other plants [11].

ALGAE CULTIVATION METHODS AND DESIGN CONSIDERATION

Algae are typically found growing in ponds, waterways or other locations that receive sunlight, water and CO_2. Manmade production of algae tends to mimic the natural environments to achieve optimal growth conditions. Growth depends on many factors and can be optimized for temperature, sunlight utilization [12, 13], pH control, fluid mechanics and more. The method behind biofixation is capturing the CO_2 and NO_x from power plant smokestacks and feeding the CO_2 to an algae system where up to 50 per cent of harmful emissions from the smokestack will be devoured. At present there are two common methods for algae based carbon sequestration: open ponds and closed photo bioreactors.

Table 3: Advantage and Disadvantages of Open and Closed Algae Growth Systems [5, 6]

Parameter	*Open Pond*	*Closed Photo Bioreactor*
Construction	Simple	More complicated
Cost	Cheaper to construct/operate	More expensive
Typical growth rates(g/m^2-day)	Low: 10-25	Variable: 1-500
Water losses	High	low
Typical biomass concentration	Low: 0.1–02 g/L	High: 2-8 g/L
Temperature control	Difficult	Easy
Species control	Difficult	Easy
Contamination	High Risk	Low Risk
Light utilisation	Poor	Very High
CO_2 losses to atmosphere	High	Almost none
Area requirements	Large	Small
Depth/Diameter of water	0.3m	0.1m
Surface to volume ratio(m^2/m^3)	~6	60-400

Open ponds are simple expanses of water recessed into the ground with some mechanism to deliver CO_2 and nutrients with paddle wheels to circulate the algae broth. Closed photo bioreactors are a broad category referring to systems that are enclosed and allowing more precise control over growth conditions and resource management. A large-scale photo bioreactor would be similar to a large display of solar panels, except instead of producing electricity, the solar energy would serve through photosynthesis by micro algae to convert CO_2 from fossil fuel combustion to

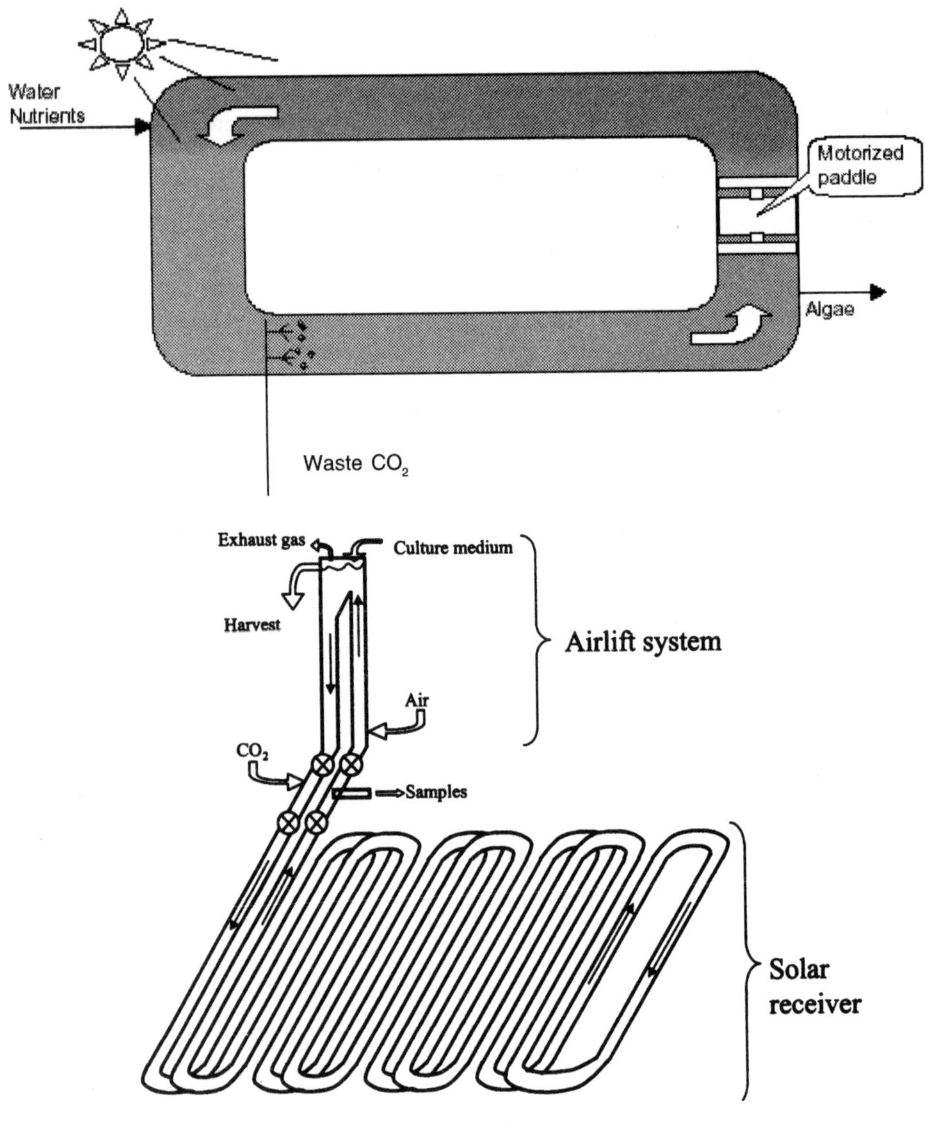

Figure 2: Algae Open Pond System and Closed Photo Bioreactors

stable carbon compounds for sequestration. Closed photo bioreactors allow more precise control over growth conditions and resource management. Each system has benefits and drawbacks with respect to optimal growth conditions. Table 3 presents a short comparison of open pond systems and closed photo bioreactors.

ALGAE HARVESTING AND OIL EXTRACTION

Production of algae is a straightforward process consisting of growing the algae by providing necessary inputs for photosynthesis, harvesting/dewatering and oil extraction. The fundamental mechanism governing algae growth is photosynthesis. It is in the photosynthesis process that light-driven reactions split water and assimilate carbon into the biomass. Energy in the form of photons is absorbed by the algae cells, which convert the inorganic compounds of CO_2 and H_2O into sugars and oxygen. The sugars are eventually converted into carbohydrates, starches, proteins and lipids within the algae cells. The basic concept of algal oil (or biofuels in general) production is to use relatively small (in total area) photo bioreactors to produce a modest amount of starter or "inoculum" culture (about 1-2 per cent of the total biomass) to seed much larger, several hundred hectares total area, open ponds. A diagram of the overall growth and harvesting process is presented in Figure 3.

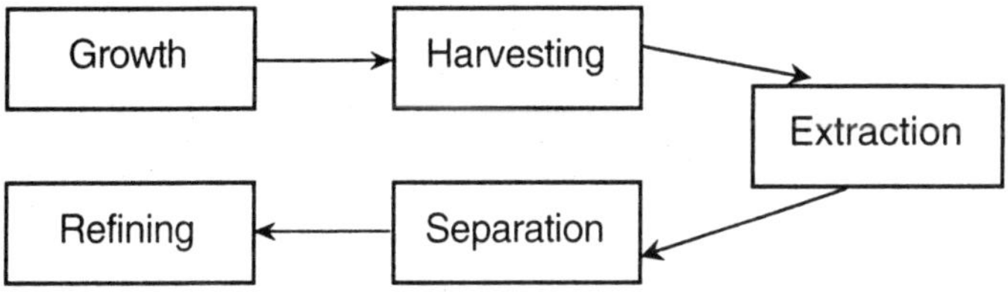

Figure 3: Algae Growth and Harvesting Process

In order to extract the valuable lipids from within the algae cells a series of steps must be undertaken to isolate the algae cells and oil. The traditional process begins by separating the algae biomass from the water broth in the dewatering stage using either centrifuges, filtration or flocculation techniques. Centrifuges collect biomass by spinning the algae-water broth so that water is flung away from the algae cells. Flocculation involves precipitating algae cells so that they can be removed out of solution. Once the algae cells have been collected the oil must be removed from the cells. Once the oil is removed it can be processed into biodiesel, jet fuel, ethanol, synthetic fuels or other chemicals. Algae for biofuels have been studied for many years for production of hydrogen, methane, oils (triglycerides for biodiesel), hydrocarbons and ethanol. Some micro algae accumulate high amounts of triglycerides when limited for nitrogen. However, the key is to produce such oils efficiently, which remains to be demonstrated. One algal species, *Haematococcus pluvialis*, contains about 50 per cent of hydrocarbons, even when growing with plenty of nitrogen. Unfortunately it grows very slowly and its mass cultivation remains a challenge.

ALGAE AS MULTIPURPOSE FEEDSTOCK

Biofuels today are primarily produced from first-generation feedstocks such as jatropha, sugarcane, soybeans and rapeseeds. Reliance on crop-based feedstocks has led to problems such as land depletion, continued fossil fuel usage, competition with food and increased water use. Algae, on the other hand, are an appealing feedstock for next-generation biofuels because they can make use of natural or underutilized resources, can be produced domestically, consume carbon dioxide via photosynthesis and have the potential to displace fossil fuel usage in an environmentally sound manner for society.

The products of algae growth can be used for many different fuels: lipids can be processed into chemical feedstocks, biodiesel or jet fuel; biomass can be fermented into ethanol, anaerobically digested to produce methane, or burned directly for power generation; or simply used as a carbon sink [14]. A diagram of beneficial uses of Algae is presented in Figure 4.

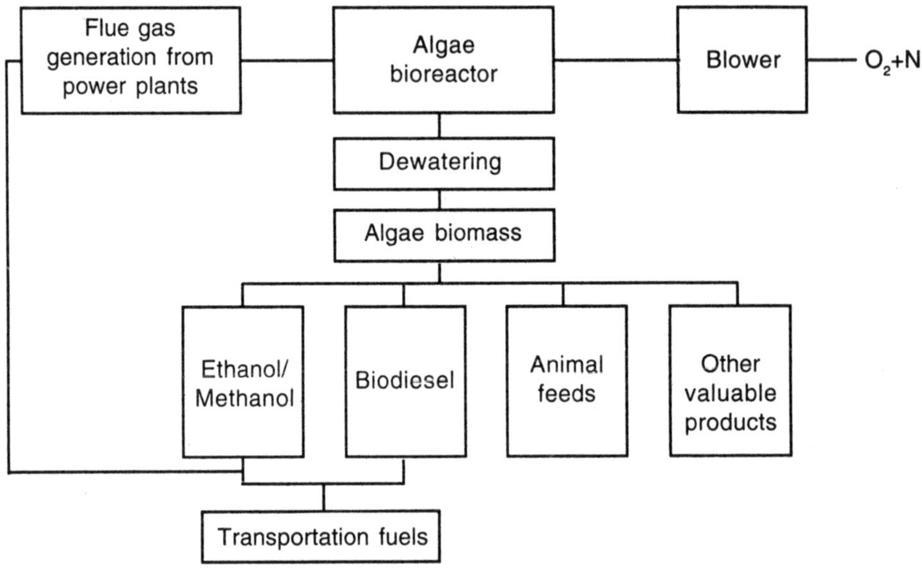

Figure 4: Beneficial Uses of Algae

CONCLUSIONS

In the endevour to reduce dependence on fossil fuels and cut carbon emissions to achieve a clean environment, humble algae appears to be taking a lead over the more-talked-about biodiesel source jatropha. Algae farming in less than 1 per cent of India's total land can make the country self-sufficient in liquid fuel. Hence algae farming for oil provides an excellent opportunity to absorb CO_2 emissions from large industrial plants and convert them to biofuel. While algae are a very promising feedstock, many challenges inhibit the production of large amounts of algae in an economic and sustainable manner. Varying levels of research is going on in labs but none have succeeded in producing algae oil on a scale sufficient for meeting our transportation requirements. The cultivation of microalgae for biofuels in general and oil production still requires relatively long-term R&D, with current emphasis on the R rather than the D. This is due in part to the high costs of even simple algae production systems (*e.g.* open paddle wheel mixed, raceway-type ponds) and in even larger part due to the undeveloped nature of the algal mass culture technologies, from the selection of suitable algal strains that can dominate in the ponds, to their low-cost harvesting and most importantly, to the achievement of the required high productivities of biomass with a high content of oil.

ACKNOWLEDGMENT

The authors are thankful to the Department of Science and Technology, Government of India for providing financial assistance to carry out this research.

REFERENCES

[1] Pirt, S.J., 1986. The thermodynamic efficiency (quantum demand) and dynamics of photosynthetic growth, The New Phytologist, 102: p. 3-37.

[2] Searchinger, T., 2008. Use of U.S. croplands for biofuels increases greenhouse gases through emissions from Land-Use Change, Science, 319: p. 1238.

[3] Haumont, D., 1993. Biotechnology of algal biomass production: a review of systems for outdoor mass culture, Journal of Applied Phycology, 5: p. 593-604.

[4] Chisti, Y., 2007. Biodiesel from microalgae, Biotechnology Advances, 25: p. 294-206.

[5] EIA, U., Petroleum Products Consumed in 2007-2008.

[6] Yeoung-Sang Yun, *et al.*, 1997. Carbon Dioxide Fixation by Algal Cultivation Using Wastewater Nutrients, Journal of Chemical Technology and Biotechnology, 69(4): p. 451-455.

[7] Cuello, E.O., 2007. Carbon Dioxide Mitigation using Thermophilic Cyanobacteria, Biosystems Engineering, 96(1): p. 129-134.

[8] Maeda, K., N. Kimura, K. Omata, I. Karubd, 1995. CO_2 fixation from the flue gas on coal-fired thermal power plant by microalgae, Energy Convers. Mgmt, 36(6-9): p.717-720.

[9] Xu, H. *et al.*, 2006. High quality biodiesel production from a microalga chlorella protothecoides by heterotrophic growth in fermenters, Journal of Biotechnology, 126: p. 499-507.

[10] Xiaoling Miaoa, Q.W., 2003. Changyan Yang, Fast pyrolysis of microalgac to produce renewable fuels, J. Anal. Appl. Pyrolysis, 71: p. 855-863.

[11] Falkowski, Paul G., Z.D., Kevin Wyman, 1985. Growth-irradiance relationships in phytoplankton, Limnol. Oceanogr, 30: p. 311-321.

[12] Perner-Nochta, I. and C. Posten, 2007, Simulations of light intensity variation in photobioreactors. Journal of Biotechnology, 131(3): p. 276-285.

[13] E. Molina Grima, F. Garcýa Camacho, F. Camacho Rubio, Y. Chisti, 2000. Scale-up of tubular photobioreactors, Journal of Applied Phycology, 12(355-368).

[14] Otto Pulz, K.S., 2001. Photobioreactors: production systems for phototrophic microorganisms, Appl Microbiol Biotechnol, 57: p. 287-293.

Chapter 8

Carbon Capture and Sequestration

D.M. Kale

Director General, ONGC Energy Centre, Scope Minar
Laxmi Nagar, New Delhi

SUMMARY

Tracking the pre-historic development of life and human evolution on earth, this paper highlights the role of carbon dioxide and need for its sequestration. Initially the CO_2 concentrations were high and earth was a fireball. As CO_2 quantity fell down the earth began to cool and became habitable. Increasing CO_2 concentrations in the atmosphere in the post-industrialization era, its stationary sources and options for sequestration are described. Whether CO_2 injection in depleting oil fields can be used for enhanced oil recovery is discussed. This area of research is in infancy and there is a need for further studies. CO_2 capture options from coal based power plants, recent developments in membrane development and geological storage techniques are also explained. Four major carbon storage projects namely; Weyburn, Sleipner, In Salah and Snohwit are currently in operation. These are expected to throw some light on the efficacy of this technology. These aspects are highlighted in the following presentation.

Guest lecture delivered in the Awareness and Capacity Building in Carbon Capture and Storage (ACBCCS-2009) Programme conducted from July 27-31, 2009 at Indian National Science Academy, Delhi.

THE HISTORY OF LIFE ON EARTH AND HUMAN EVOLUTION

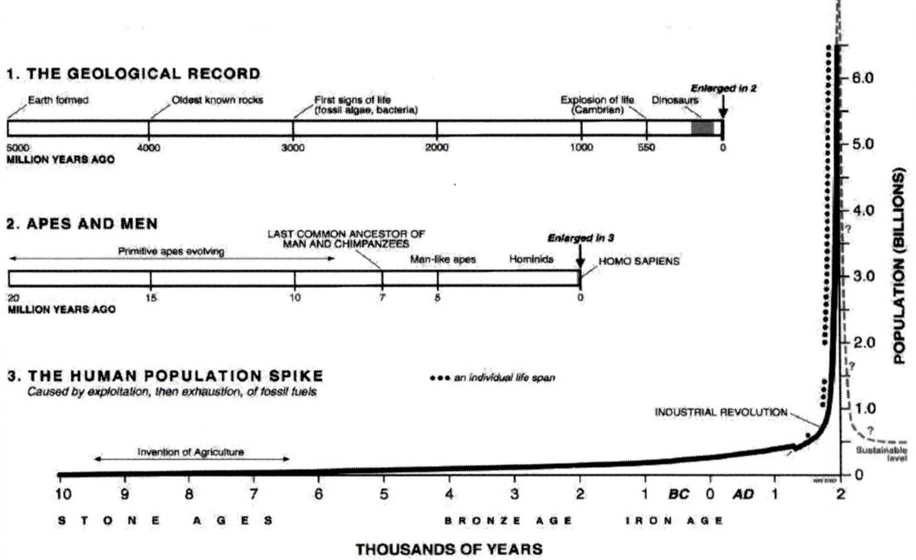

THE MIGRATION HISTORY OF HUMANS: DNA STUDY TRACES HUMAN ORIGINS ACROSS THE CONTINENTS

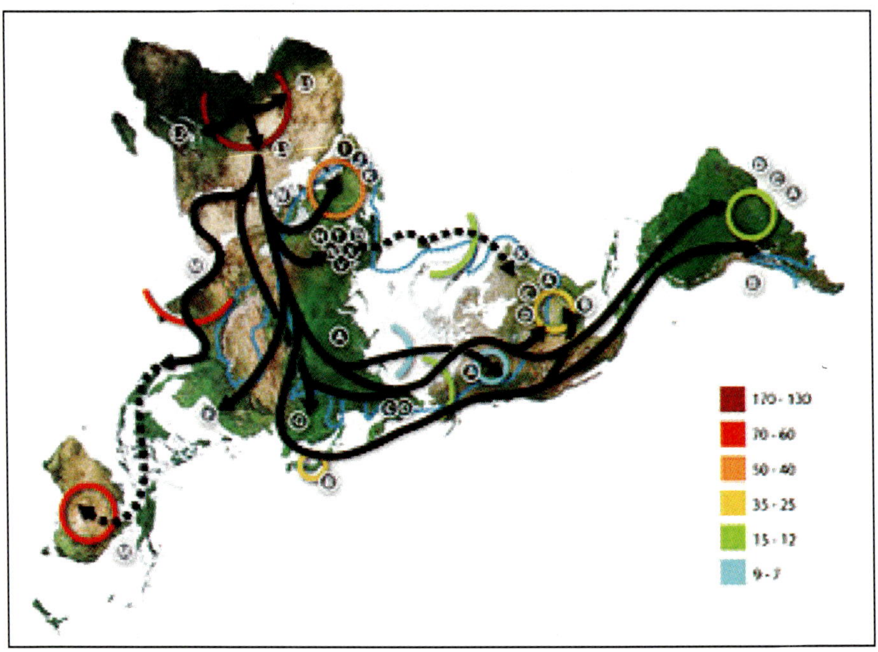

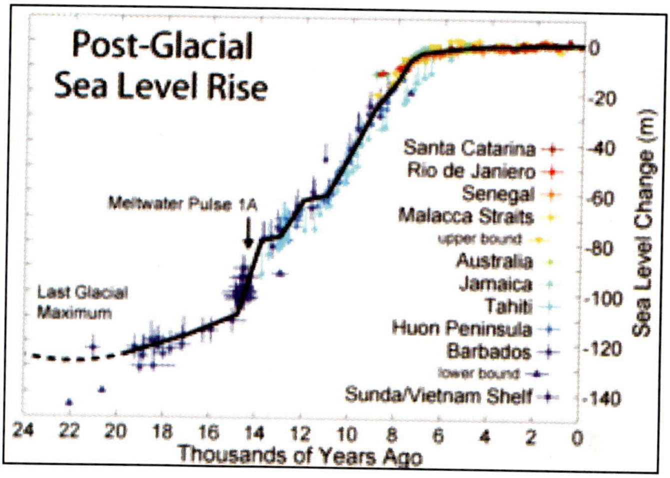

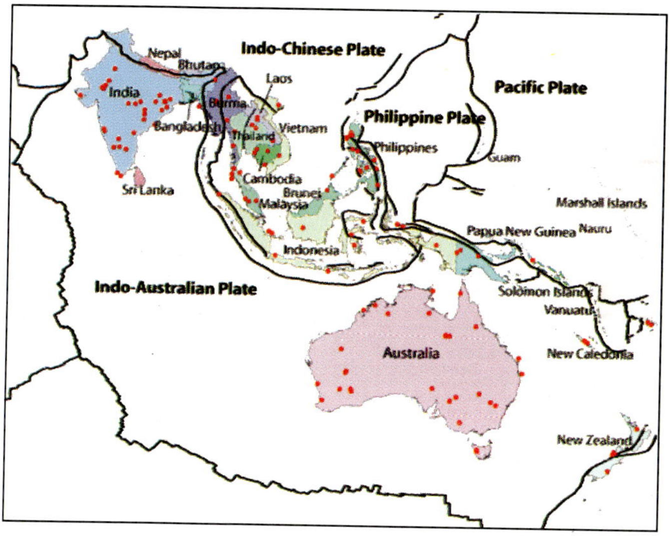

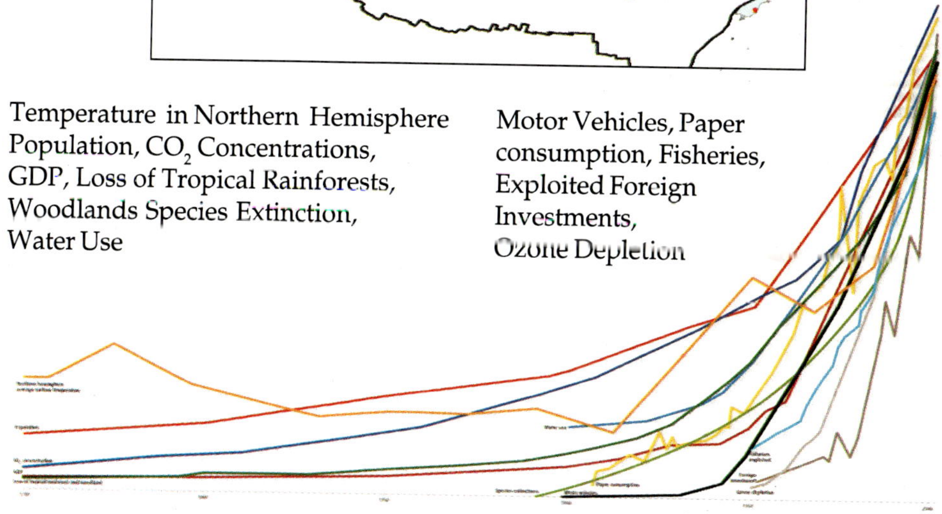

*Map not to scale

POPULATION - DRIVEN ENERGY DEMAND

World Population (Millions)

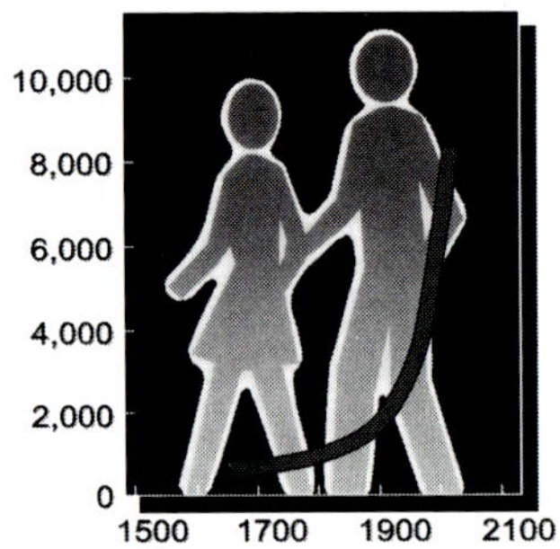

World Primary Energy Consumption (Quadrillium BTU)

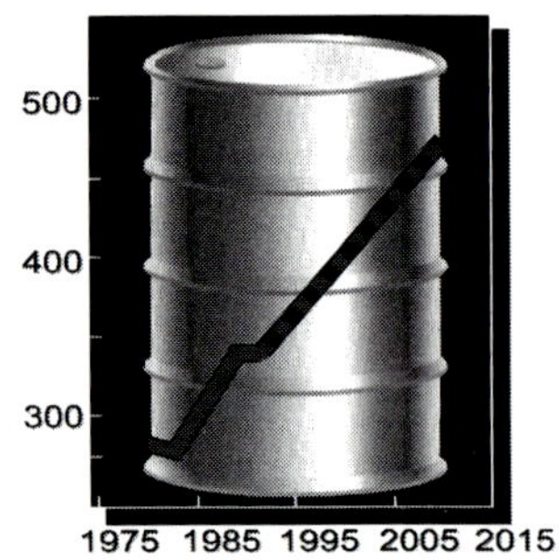

CO_2 Concentrations in the Atmosphere

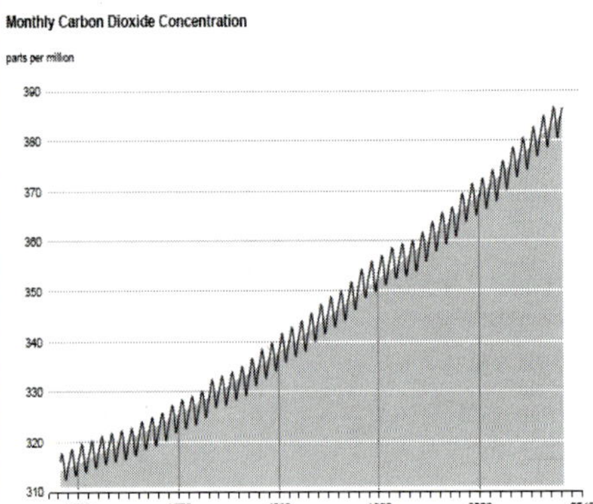

CO_2 Absorption Spectra

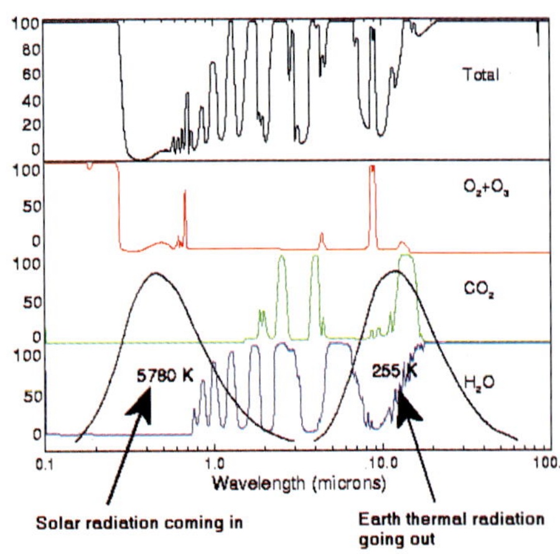

BE QUANTITATIVE: GOD IS IN DETAILS AND DEVIL IN NUMBERS!

☆ 1 ppm is 8.08 billion tons CO_2.

☆ Current CO_2 emission ~ 30 billion tons
that amounts to 3.7 ppm
of which 57% = 2.1ppm
retained in the atmosphere

- The current CO_2 emissions ~ 28 Gt
- Business as usual projection to 2050 ~ 62 Gt
- ACT plan 2050 ~ 2005 reduction by 35 Gt
- BLUE plan 2050 ~ half of 2005 reduction by 48 Gt
- Of this about 20% reduction to be achieved by CCS
- 7 to 10 GT CO_2 to be sequestered every year!

☆ **Various Simultaneous Methods for Emission Reduction**
- ☆ Renewable
- ☆ End use efficiency (fuel, power)
- ☆ Fuel Switch
- ☆ Nuclear
- ☆ CCS Power
- ☆ CCS Industry

CAUSES AND REMEDIES

Sources - Stationary

- ☆ Power Plants
- ☆ Transport
- ☆ Industry
- ☆ Cement, Iron and Steel
- ☆ Petrochemicals
- ☆ Natural Gas, Refinery
- ☆ Residential

Sources - Mobile

- ☆ Vehicular pollution
- ☆ A few Billion of mobile polluters wheezing all over the place !

- 69% of CO_2 emission and 60% of GHG emission Energy Related!
- 2005 Electricity Consumption 18000TWh
- Projected to increase by 179% to 50000TWh

Remedies - Sourcing electricity from

- ☆ Solar
- ☆ Wind
- ☆ Biomass
- ☆ Biomass CCS
- ☆ Hydro
- ☆ Nuclear
- ☆ Gas
- ☆ Gas CCS
- ☆ Coal
- ☆ Coal CCS

LOOK AT THE PROBLEM FROM ALL POSSIBLE ANGLES!

Carbon Capture and Sequestration Potential for CO_2 EOR ~200 billion barrels of extra oil will lead to storage of 70 to 100 Gt

- ☆ Enhanced Oil Recovery
- ☆ Miscible displacement
- ☆ Minimum Miscibility pressure
- ☆ Immiscible displacement
- ☆ Little Incremental Recovery
- ☆ 3% of total oil production from EOR
- ☆ 10% of that from CO_2
- ☆ 0.3% of total oil production by CO_2

This is about the CO_2 emission from Oil! Do we do any net good!

CO_2 Production per Tonne

Source	*Amount in kg*
Conventional oil production	150
Refining	180
LNG	200
Heavy Oil production	410
Heavy oil upgradation	750
GTL	1150

CO_2 Intensity of Power Plants

In gms / kWh	*Now*	*In 2050*
Non OECD	600	90
OECD	450	40
Average	525	70

CO_2 CONTRIBUTION BY VARIOUS SOURCES

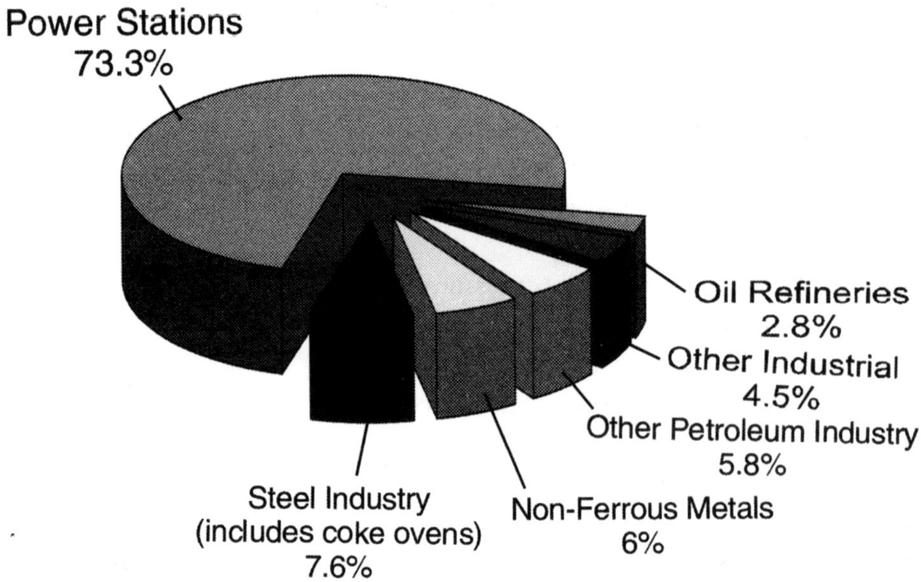

GEO-SEQUESTRATION: Carbon Capture and Storage

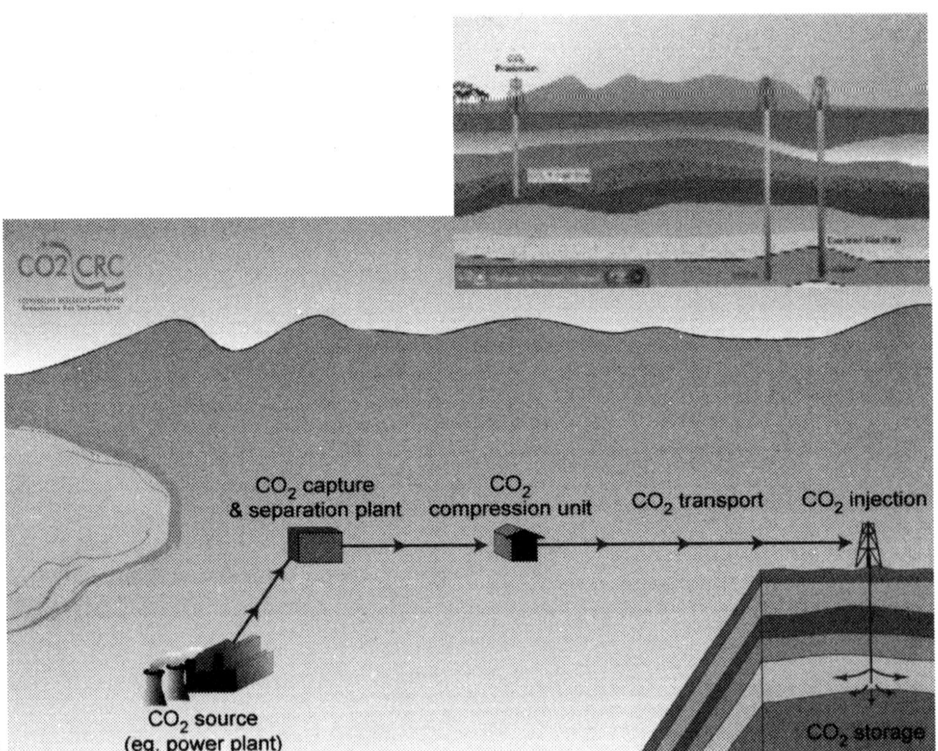

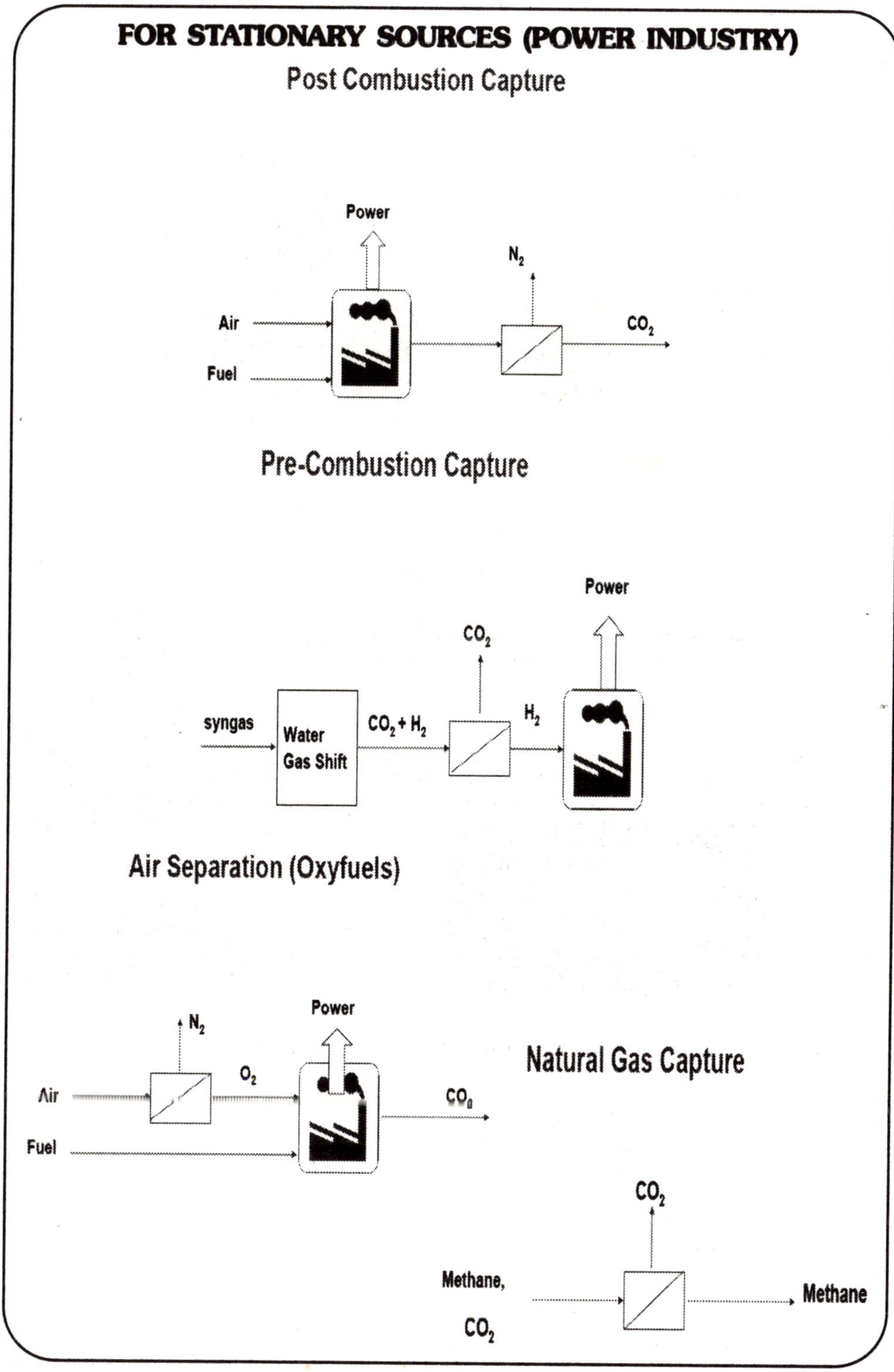
FOR STATIONARY SOURCES (POWER INDUSTRY)
Post Combustion Capture
Power
N_2
Air
Fuel
CO_2
Pre-Combustion Capture
Power
CO_2
syngas
Water Gas Shift
$CO_2 + H_2$
H_2
Air Separation (Oxyfuels)
N_2
Power
O_2
Air
Fuel
CO_2
Natural Gas Capture
CO_2
Methane, CO_2
Methane

ADSORBENTS AND MEMBRANES

Zeolites

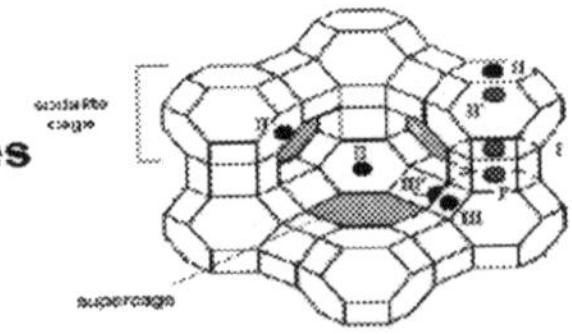

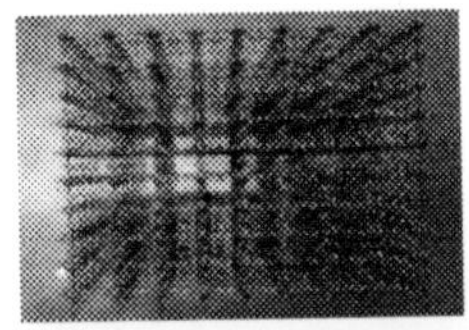

Metal Organic Frameworks (MOF)

Mesoporous Carbons

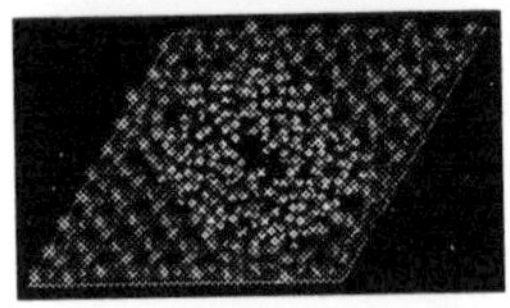

Inorganic-Organic Hybrids

Within each class, there are many structural variations.
Each class/variation has advantages and disadvantages.
Specific variations are appropriate to different process configurations.

Membranes

CO_2 can selectively pass through gas separation membranes to be extracted from the flue gas.

Membrane/solvent

A membrane separates flue gas from liquid solvent. CO_2 is absorbed by the solvent via pores in the membrane.

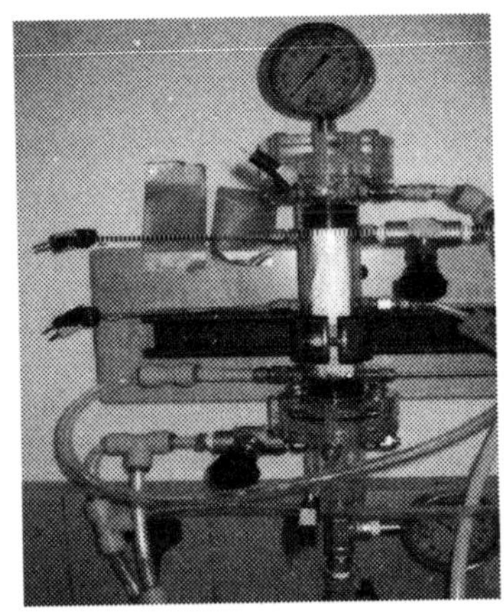

Membranes for CO_2 removal

- First Generation:
  - Cellulose acetate (Cynara, UOP, Grace)
  - Polysulfone (Air Products' Prism)

  Generally spiral wound
- Second Generation
  - Polyimides (Ube, MEDAL)

  More likely to be hollow fibre

CARBON DIOXIDE: PRESSURE-TEMPERATURE PHASE DIAGRAMS

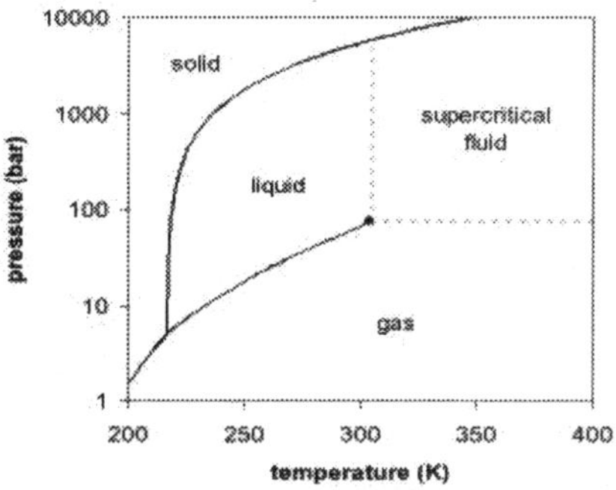

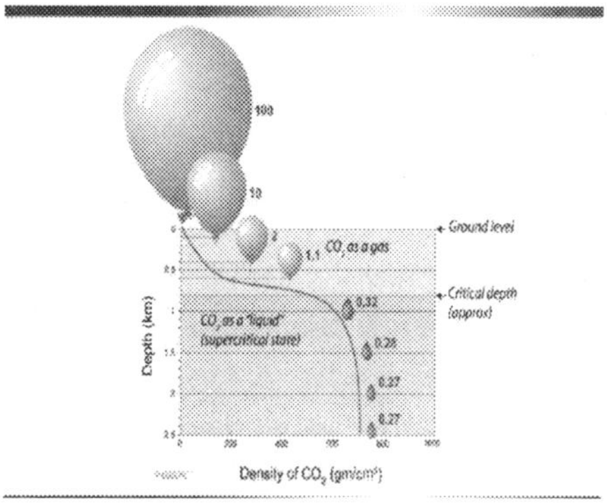

Geometry of injected CO_2 'bubbles'

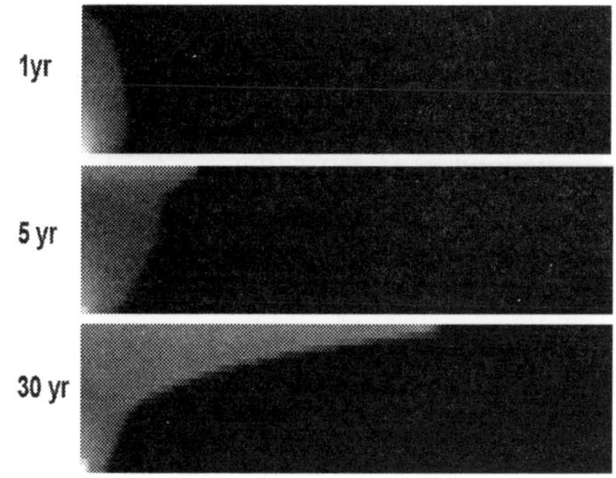

Geometry of injected CO_2 'bubbles'

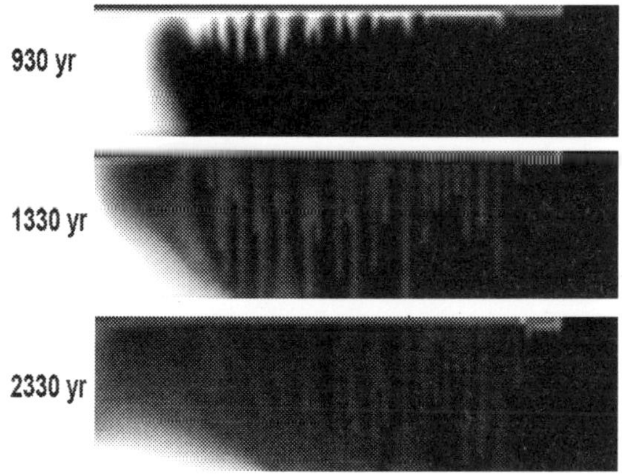

CAPACITY OF CARBON STORAGE

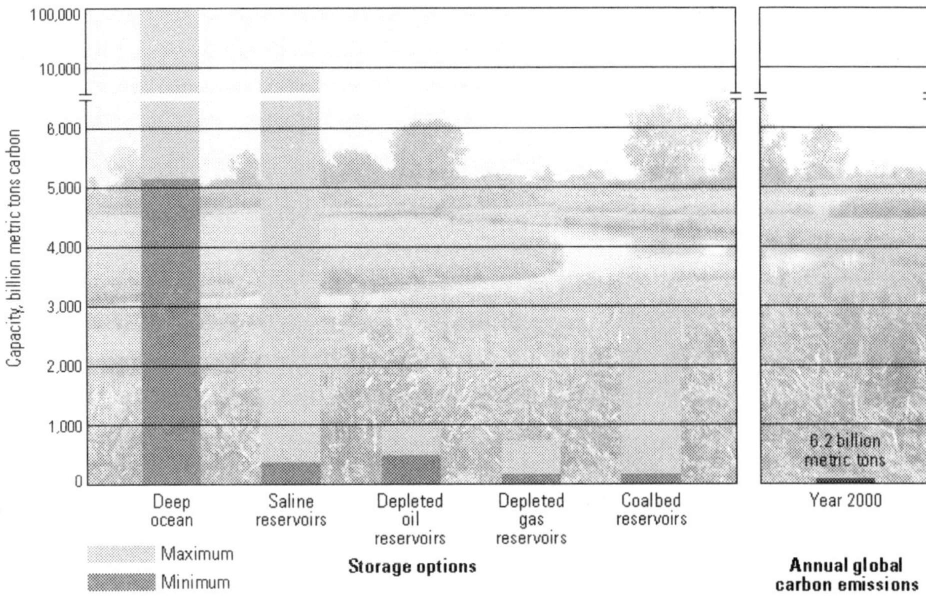

CO_2 STORAGE PROJECTS - CURRENT & PROPOSED

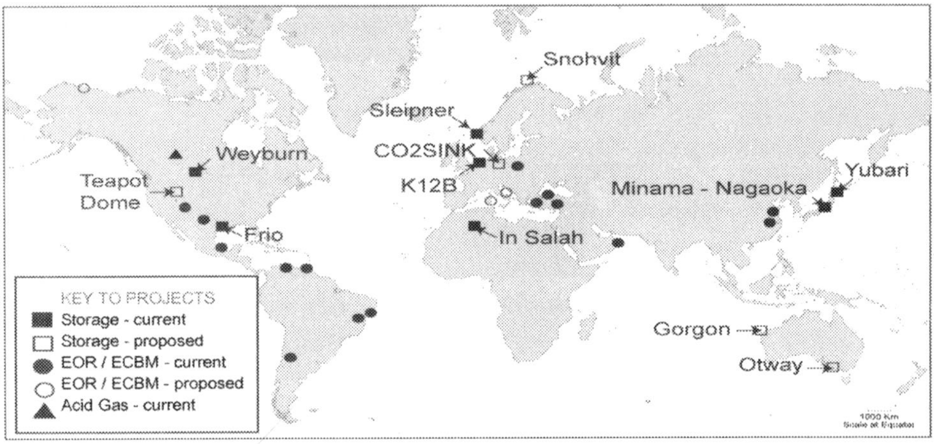

CO_2 SEQUESTRATION PROJECTS

Sleipner (STATOIL)

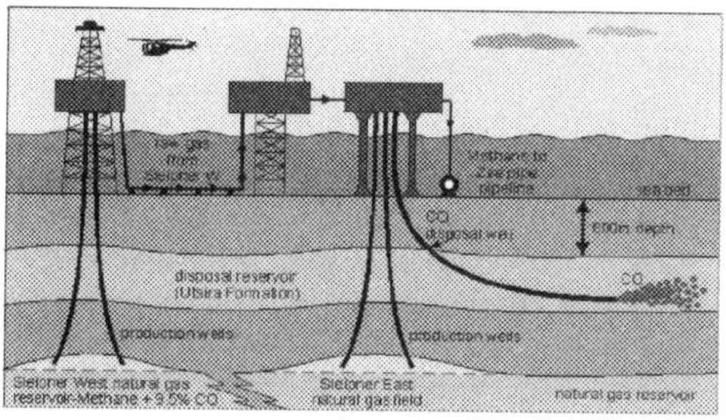

- ☆ 250 kilometers west of Norway in the North Sea
- ☆ Injection into Utsira Formation, a sandstone
- ☆ 1 million tons of CO_2 per year since 1996

Weyburn CO_2 Project

CO_2 Source: Dakota Gasification Company
95 mmscfd (5000 tonnes/day) injection rate
CO_2 purity 95% (primary feed)
Currently 26% recycle.

Main CO_2 pipeline enters Weyburn

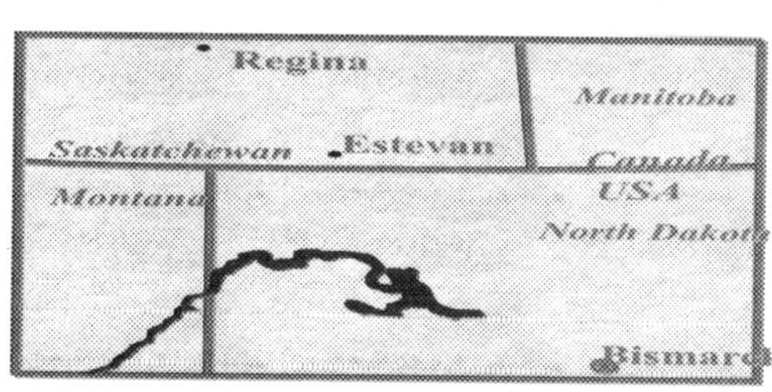

CO_2-SEQUESTRATION PROJECTS

Contd...

In Salah CO_2 Storage

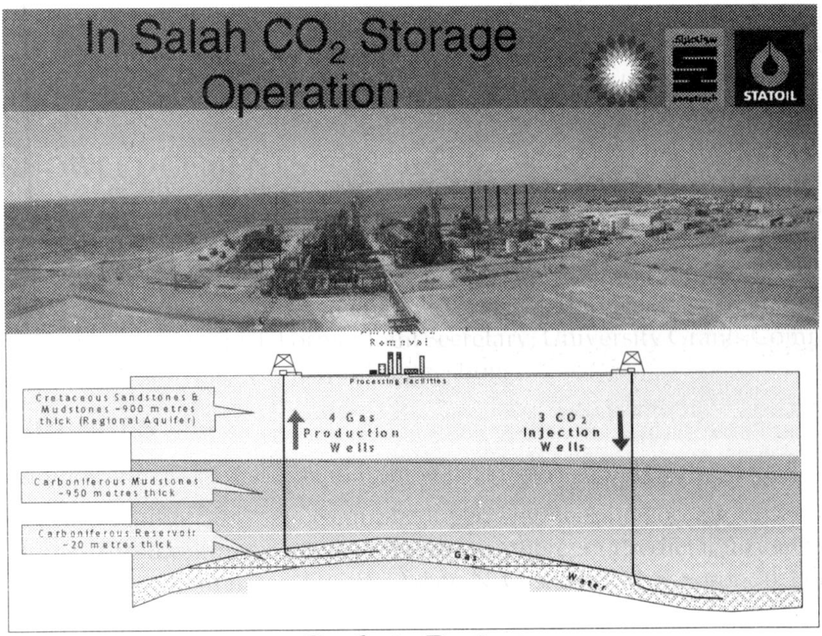

Snøhvit Project

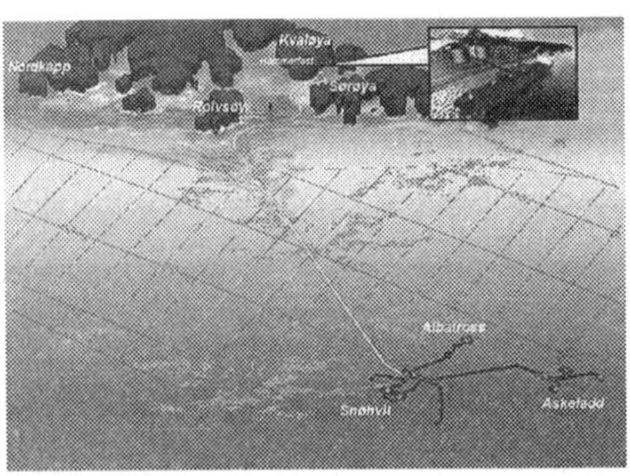

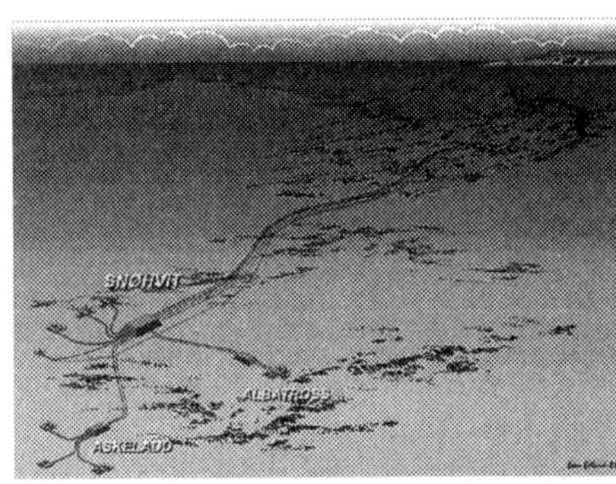

- ☆ In April 2008, Statoil announced carbon storage had started on its Snøhvit field – Statoil is reinjecting Snøhvit's CO_2 emissions into the ground beneath the gas-bearing formation on the field. The process will reduce CO_2 emissions by 700,000t a year when Snøhvit is at full capacity, it is estimated. This is the equivalent of emissions from 280,000 cars.

- ☆ Natural gas is first pumped to a carbon capture plant at Melkøya. Here, 5% to 8% of CO_2 is removed from the gas and piped back to a 2,600m deep sandstone formation at Snøhvit, where it sits under the seabed.

INDIA CO_2 SOURCES & STORAGE POTENTIAL– SOME CONCERNS

- ☆ Third largest coal user
- ☆ 62% Energy derived from coal
- ☆ 75% of coal used for power generation
- ☆ Banks on coal for future needs 9 ultra mega power projects to add 36 GW

- ☆ Coal has high ash complicates capture.
- ☆ IGCC capture more expensive.
- ☆ TWO plants for capture of CO_2 450 t /day in Fertilizer industry at Aonla, Phulpur

- ☆ Storage Potential: 500 to 1000 Gt
- ☆ Off shore Deep Saline: 300 – 500 Gt
- ☆ Basalt Traps: 200 – 400 Gt
- ☆ Depleted Oil, Gas fields: 5 --10 Gt
- ☆ Unminable Coal Seams: 5 Gt

- ☆ No field large enough to store life time emissions from a medium sized power plant
- ☆ Deccan Volcanic Province 500,000 km^2 require further study
- ☆ Saline Aquifer potential in Assam more than 1000 km away from CO_2 sources

Chapter 9

Role of Algae in Carbon Sequestration

Dinabandhu Sahoo

Marine Biotechnology Laboratory, Department of Botany, University of Delhi, Delhi – 110 007

SUMMARY

Today global warming is a major concern all over the world, which is caused due to emissions of high amount of greenhouse gases which include carbon dioxide, methane, nitrous oxide etc. Out of these gases, CO_2 is of major concern. The level of CO_2 in the atmosphere has increased by 31 per cent since 1975 and presently CO_2 emissions are 7.4 billions tons per year which may increase to 20 billion tons by 2100. Such an enormous increase of CO_2 in the atmosphere will have disastrous environmental consequences. So, there is the necessity to not only control the CO_2 emissions but also to reduce it by 25 per cent. There are several ways by which the CO_2 can be reduced from the atmosphere. Presently out of 7.4 billions of CO_2 emissions terrestrial-biosphere sequesters 2 billions where as oceanic uptake sequesters 2 ± 0.8 billion tons. Algae, one of the most important living resources of the oceans can be used for CO_2 sequestration in an efficient manner. It has been found that 1 ton of algae can fix 0.36 tons of carbon, 0.6 tons of nitrogen and 0.008 tons of phosphorus. It has been also found that annual carbon fixation by intertidal marine algae in the estuaries exceeds 13,500 tons per year and account for 21 per cent of total carbon fixed by all primary producers. Although, marine algae popularly known as seaweeds are being used as a source of food, fodder, medicine and various fine chemicals, its role in carbon sequestration has never highlighted. Over the last 50 years, although seaweeds cultivation has been practiced in several oriental countries, its environmental benefit has not been realized to the full extent. Mariculture of seaweeds is now expanding in several tropical countries. India has a long coastline of 7000 km with 717 species of seaweeds. Till recently the seaweeds are mainly harvested from natural resources for commercial use. Recently, large scale cultivation of seaweeds such as Kappaphycus alvarezii and Gracilaria verrucosa has started in different parts of the Indian coast line. Efforts are now been initiated to expand the seaweed cultivation to several other parts of the Indian coasts. Details of seaweed cultivation, its application and role in carbon sequestration are discussed.

Six global risks key to India's future

- Water shortage
- Oil shocks-Energy Deficit
- Mass unemployment
- Food and nutritional security
- HIV/ AIDS
- Climate change- Global Warming

MICRO ALGAL SOURCE

30,000 Micro Algal species.
3-4-Commercially exploited

Year	Species	Product
1960's	*Chlorella*	*B*-Carotene
1970's	*Spirulina platensis(* Arthospira)	
1980's	*Dunaliella salina*	*B-Carotene Provitamin A*
1990's	*Haematococcus pluvialis(Chlorophyta)*	*Astaxanthin*
2000's	*Porphyridium(Rhodophyta*	Sulphated Polysaccharide
	Nanochloropsis	*Poly unsaturated fatty acid*
		B-Carotene, Astaxanthin
		Astaxanthin

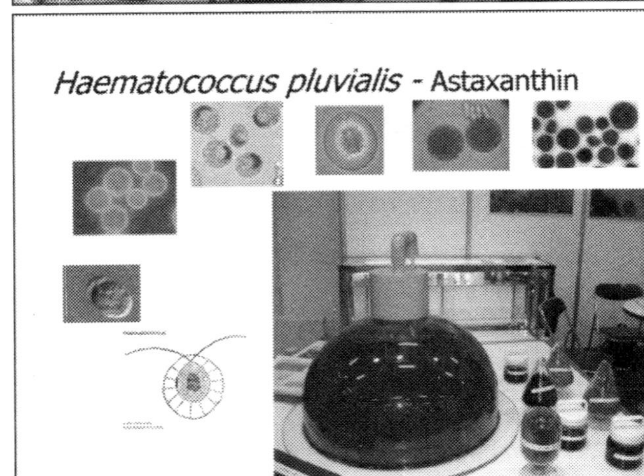

Commercial cultivation of *Haematococcus*

➢There are nearly 200 companies developing algae as Biodiesel feedstock.

➢Out of these 35 viable companies are working in the field.

➢When, the price rise is $ 1/ Barrel pentagon's fuel expenses climb to $ 130 million.

➢During the past six months the oil prices has increased by more than $ 50/ barrel causing a loss of half a billion dollar to pentagon.

➢US consume 20.7 million barrels per day (bpd) of oil which includes an import of 12.4 million barrels per day

➢Since food crops are diverted for biofuels, these first generation biofuels are called AXIS OF EVIL.

➢Soya oil prices have risen more than 35% in the past six months.

ALGAL BIOFUEL COMPANIES

NAME	COUNTRY	FINANCIAL BACKUP	CAPITAL INVESTMENT	TIEUPS / COLLABORATER
Green fuel technologies	Cambridge, Masschusets ,USA	•Polaris Ventures •Access private equity •Draper Fisher Jurvetson	$ 92@ million	
Solazyme	San Fransisco, USA	•Blue Crest Capital Finance •The Roda Group		•Chevron, • Imperium Renewables (a biodiesel maker)
Cellena	KonaCoast, Hawai, USA			•HR, Biopetroleum, •Shell
Algae @work (A2BE)	Colorado,USA			
Petrosoun	Rio Hondo,Texas, USA			
Solazyme	San Fransisco, USA	•Rhoda group •Hurris & Hurris Group •Braamer Energy Venture •Lightspeed Venture poreners	$ 45 million	Chevron

List of algae with oil contents

Microalga	Oil content (% dry wt)
Botryococcus braunii	25-75
Chlorella protothecoides	55
Chlorella vulgaris	14-22
Scenedesmus dimorphus	16-40
Isochrysis sp.	25-33
Nannochloris sp.	20-35
Nannochloropsis sp.	31-68
Neochloris oleoabundans	53-54
Nitzschia sp.	45-47
Prymnesium parvum	22-38
Tetraselmis chui	45.7
Schizochytrium sp.	50-77
Scenedesmus sp.	12-40
Prymnesium sp.	22-38

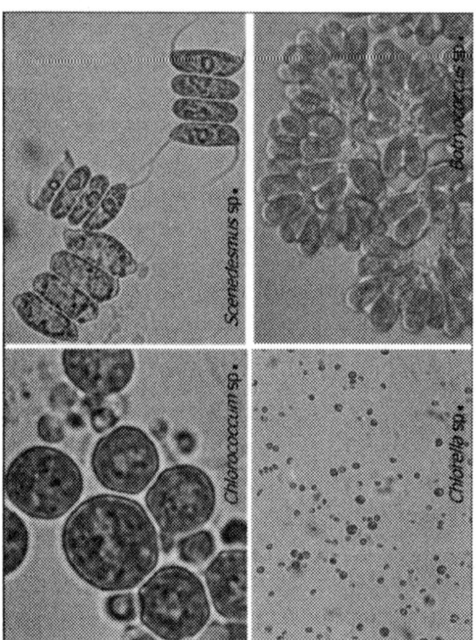
Chlorococcum sp.
Scenedesmus sp.
Chlorella sp.
Botryococcus sp.

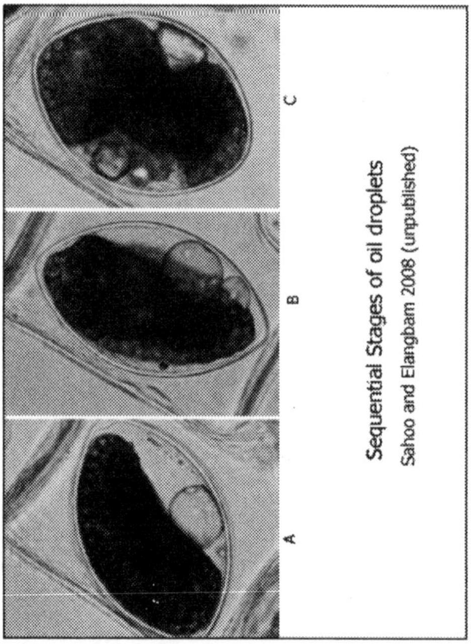
A
B
C
Sequential Stages of oil droplets
Sahoo and Elangbam 2008 (unpublished)

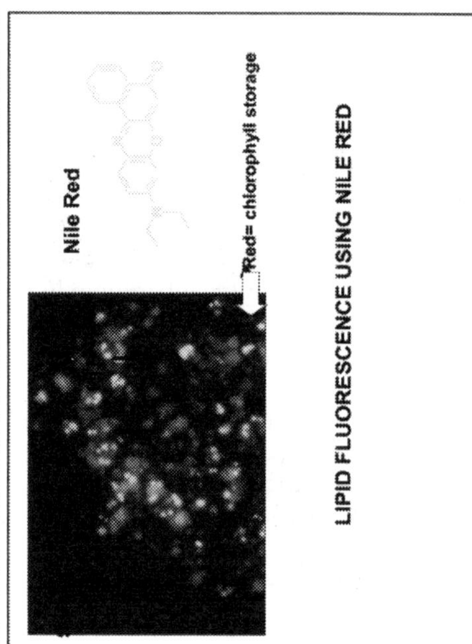
Nile Red
Red= chlorophyll storage
LIPID FLUORESCENCE USING NILE RED

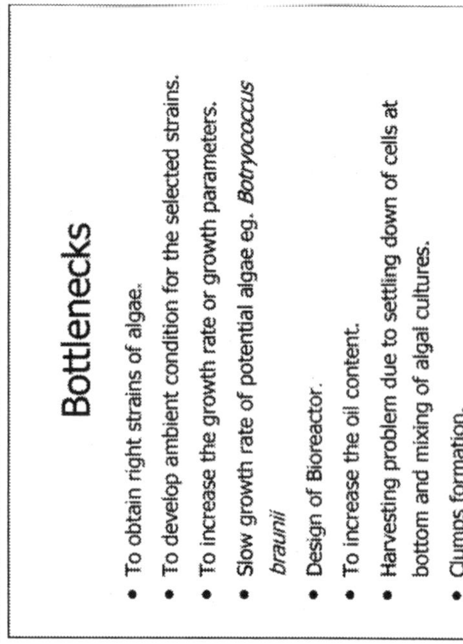
Bottlenecks
To obtain right strains of algae.
To develop ambient condition for the selected strains.
To increase the growth rate or growth parameters.
Slow growth rate of potential algae eg. Botryococcus braunii
Design of Bioreactor.
To increase the oil content.
Harvesting problem due to settling down of cells at bottom and mixing of algal cultures.
Clumps formation.

INDIA'S POSITION

770 Species of Seaweeds (Sahoo et al., 2001)

Large coast line 7000 km

National Coordinated Program for large scale cultivation and utilization of 3-4 taxa having both domestic and international market.

Gracilaria verrucosa

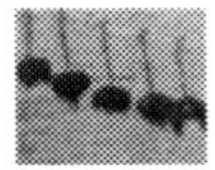

Kappaphycus alvarezii

Sargassum sp.

Porphyra sp.

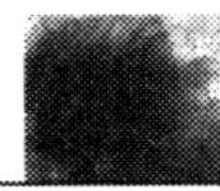

Comparative chart of some oil yielding microalgae

Microalgae	Habitat	Total Lipid content (%dry weight)	Growth rate(d^{-1})	Doubling time(day)	Harvesting time
Botryococcus braunii	Freshwater, brackish water	25-75	1.25	2 7(hydrogen production)	
Ankistrodismus sp.		40.3	2		
Chlorella sp.	**freshwater**	**44**		1	
Nannochloropsis sp.	Marine	31-68			
Scenedesmus dimorphus	Freshwater, Sewage pollution	16-40			

Botryococcus braunii in different geographical area (Metzger.P et.al, 2005)

	Bolivia	UK	Morocco	Martinique	Ivory Coast
Total lipid (% dry wt.)	62	63	43	53	35

Role of seaweeds in Global Warming

The level of CO_2 in earth's atmosphere has increased by 31% since 1750.

•1997 -- CO_2 emission 7.4 billion tons/year.

• 2050 – CO_2 emission will be 15 billion tons/year.

•Recently it has been realized that seaweeds also play a major role in carbon sequestration.

•1 ton of seaweed can fix 0.36 tons of carbon, 0.6 tons of nitrogen and 0.008 tons of phosphorus.

•Annual carbon fixation by inter-tidal seaweeds alone in the estuaries exceeds 13,500 tons carbon per year and accounts for 21 % of total carbon fixed by all primary producers.

Chapter 10

Enzyme and Microbe Mediated Carbon Sequestration

Adarsh K. Puri and T. Satyanarayana
Department of Microbiology, University of Delhi, South Campus, New Delhi – 110 021

SUMMARY

The concentration of atmospheric CO_2 has increased in an unprecedented manner after the industrial revolution. The whole scientific community, politicians and industrialists throughout the world are debating on developing strategies to bring down the atmospheric carbon levels to the acceptable limits. Carbon sequestration has emerged as the most potential and effective way for mitigating global warming in the past few years. Carbon capture and storage in ocean and deep geological formations are the promising solutions. The current focus now, however, is on novel ideas that ensure a leakage-proof and cost-effective approach for long-term and maximal storage of CO_2. Capturing carbon by biological means is not only a means of sequestering carbon, but may also lead to the production of useful products. A proper understanding of enzymes and heterotrophic microbial systems would help in stabilizing atmospheric carbon through photoautotrophic and non-photosynthetic CO_2 fixation processes. The role of carbonic anhydrase as biocatalyst and biomimetic processes in CO_2 sequestration are explained.

INTRODUCTION

The thawing permafrost, melting glaciers, rising sea levels, changing hydrological cycle, increasing precipitation, declining crop productivity, early breeding of birds, vanishing coral reefs, increasing health hazards are all predictable effects of the hottest debatable phenomenon on earth, known as global warming. This is frequently referred to as climate change, which is not just a theory or a distant threat. The 2007 Nobel peace prize to Intergovernmental Panel on Climate Change (IPCC) and Albert Arnold Al Gore has made every one realize the severity of the problem, and further cleared all doubts being raised by naysayers over its reality. The overwhelming agreement among the world's prominent scientists, governments and scientific bodies is that the Earth is heating up and that human activities are largely to blame. The global warming is expected to significantly disrupt the planet's climate system. Minimization of greenhouse gas emissions to acceptable limits is the intrinsic environmental responsibility of the whole world.

Over the last 200 years since the Industrial Revolution, most of the world's energy has been derived from burning the finite resources of fossil fuels, mainly coal, oil, and more recently gas. Fossil fuels account for 80 per cent of the global energy demand. During the process, billions of tons of carbon dioxide and other greenhouse gases (GHGs) have been spewed into the atmosphere. Energy sector accounts for the greatest share (36 per cent) of carbon dioxide emissions. A large 1000 megawatt coal power station releases around 5.5 million tons of CO_2 annually.

Earth's atmosphere is essentially transparent to incoming radiation from the Sun, as sunlight peaks in the visible part of the spectrum. On the other hand, thermal radiation from the Earth, in the form of long-wavelength infrared rays, lies in the absorption spectrum of carbon dioxide and other GHGs. The GHGs absorb radiation primarily in a very narrow frequency band (7-13μm), while CO_2 absorbs over a much larger (13-19μm) spectral range. That is why CO_2 accounts for higher (21 per cent) greenhouse effect (after water vapour that accounts for 64 per cent) than ozone (6 per cent) and other trace gases (9 per cent).

Carbon dioxide makes up 68 per cent of the total greenhouse gas emissions. The atmospheric CO_2 concentration has increased from 280 ppm in 1800, the beginning of industrial age, to 390 ppm today. Without any mitigation, it could reach levels of 700-900 ppm by the end of the 21st century, which could bring about severe climate change. The annual CO_2 concentration growth rate was larger during the last 10 years (1995-2005 average: 1.9 ppm per year) than it has been since the beginning of direct atmospheric measurements.

This abrupt imbalance has disturbed the Earth's carbon cycle that is normally kept in balance by the oceans, vegetation, soil and the forests. The most pressing technical and economic challenge of the present time is to supply energy demand for the world economic growth without affecting the Earth's climate. That is why the current focus is on reducing fossil fuel usage and minimizing the emission of CO_2 in atmosphere. In spite of the great advances made in the field of renewable energy, it has not been possible to replace gas, coal and oil to meet the current energy needs. If fossil fuels, particularly coal remain the dominant energy source of the 21st century, stabilization of the concentration of atmospheric CO_2 will require development of the capability to capture CO_2 from the combustion of fossil fuels and store it safely away from the atmosphere. The hazards of global warming have reached to a magnitude that irreversible changes in the functioning of the planet are seriously endangered. It is, therefore, implacable for the whole scientific community to restore permissible levels of CO_2 by using the existing knowledge.

Carbon sequestration or carbon capture and storage (CCS) has emerged as a potentially promising technology to deal with the problem of global warming. Several approaches are being considered, including geological, oceanic, and terrestrial sequestration, as well as CO_2 conversion into useful materials. In this chapter, an attempt has been made to review the possible strategies for carbon sequestration.

GASES CONTRIBUTING TO GLOBAL WARMING: GREENHOUSE GASES

Greenhouse gases trap the heat that is expected to escape from earth. The extent of greenhouse effect contributed by different gases over a certain timeframe is expressed in terms of their individual Global Warming Potential (GWP) taking CO_2 as the reference gas.

The main greenhouse gases produced by human activity are carbon dioxide (CO_2), methane (CH_4), nitrous oxide (N_2O) and some halogenated compounds with high GWP. Perfluorocarbons (PFCs), sulphur hexafluoride (SF_6) and hydrofluorocarbons (HFCs) were added to the list of greenhouse gases under the Kyoto Protocol to the United Nations Framework Convention on Climate Change (UNFCCC) in 1997. Non-CO_2 greenhouse gases are also a matter of concern owing to their significant contribution (~30 per cent) to the overall anthropogenic greenhouse effect. The amount of anthropogenic CO_2 emitted to the atmosphere is much greater than any of other greenhouse gases. As a result, CO_2 makes the highest contribution to the greenhouse effect despite its low GWP.

CARBON SEQUESTRATION AND ITS IMPORTANCE

Carbon sequestration can be defined as the capture and secure storage of carbon that would otherwise be emitted to or remain in the atmosphere. The idea is to keep carbon emission produced by human activities from reaching the atmosphere by capturing and diverting it to secure storage. Much work, however, remains to be done to understand the science and engineering aspects and potential of carbon sequestration options. Given the magnitude of carbon emission reductions needed to stabilize the atmospheric CO_2 concentration, multiple approaches to carbon management will be needed. The natural carbon cycle is balanced over a long term, but is dynamic over the short-term. Historically, acceleration of natural processes that emit CO_2 is eventually balanced by the acceleration of processes that sequester it, and vice versa. The current increase in atmospheric carbon is the result of anthropogenic mining and burning of fossil fuels. Developing new sequestration techniques and accelerating existing techniques would help in diminishing the net positive atmospheric carbon flux.

METHODS OF OCEAN AND GEOLOGICAL CARBON SEQUESTRATION

Oceans cover over 70 per cent of the Earth's surface with an average depth of about 3800 metres. Depending upon the oceanic equilibrium with the atmosphere, a significant amount of captured CO_2 could be deliberately injected into the ocean at great depth, where it would remain isolated from the atmosphere for centuries. Direct ocean CO_2 disposal is now the biggest hope to use ocean as the largest sink for carbon sequestration purposes. Disposal of CO_2 directly into the oceans is discussed below.

The research on ocean disposal options has mostly focused on predicting the behavior and the dissolution time scale of the released CO_2. Different scenarios of CO_2 disposal in the ocean have been proposed at various depths and in different forms in relation to the phase properties of CO_2. CO_2 can be released directly into the ocean in any of its physical forms- gas, liquid, solid or solid hydrate. It is, however, important to study CO_2 induced density changes on the fluid dynamics of the ocean before its release. Dissolved CO_2 increases the density of seawater. that affects its transport and mixing. Density of injected CO_2 is also controlled by geothermal gradient, which varies from 0.02°C/m to 0.04°C/m. The rate of CO_2 dissolution in the seawater depends upon its physical form (gas, liquid, solid or solid hydrate), the depth and temperature of disposal, and the local water velocities.

CO_2 could potentially be released as a gas above 500m depth. However, due to lesser density of gas bubbles than surrounding seawater, these bubbles tend to rise up on the surface, dissolving at a radial speed of about 0.1 cm hr^{-1}. It is better to use CO_2 diffusers to produce smaller CO_2 bubbles, which can dissolve completely before reaching the surface.

When liquid CO_2 is injected into a sea floor depression at a depth greater than 3,000 m (where it is denser than sea water), it accumulates as a stable large "lake" of CO_2. The dissolution of these liquid CO_2 lakes is retarded by formation of a thin hydrate layer over it. While investigating different kinds of discharge pipes for CO_2 lake creation on sea floor, Nakashiki in 1997 proposed a 'floating discharge pipe' that was simple and less likely to be damaged by wind and wave in storm conditions. Slurry of liquid CO_2 mixed with dry ice in discharge pipe provides good conditions for lake formation. Oceans can sequester so much of CO_2 not only because of their large volume but also because CO_2 dissolves in water to form various ionic species that increase the total dissolved inorganic carbon (DIC) of seawater. Total dissolved inorganic carbon (DIC) is the sum of carbon contained in H_2CO_3, HCO_3^- and CO_3^{-2}.

$$CO_2(g) + H_2O \longleftrightarrow H_2CO_3\,(aq.) \longleftrightarrow HCO_3^- + H^+ \longleftrightarrow CO_3^{-2} + 2H^+$$

Ocean surface water is supersaturated with respect to calcium carbonate, while the deeper ocean water would be with lower pH and remain under saturated. This makes organisms to produce calcium carbonate particles (*e.g.* corals) in the surface oceans, which settle and dissolve in under saturated regions of deep oceans.

Since the first use of CO_2 for large-scale recovery of residual oil from Texas reservoirs in 1972, the concept of using CO_2 for beneficial purposes has got momentum.

Long-term operational experience with geological formations, its substantial capacity as a CO_2 sink, and its immediate availability has led to consideration of global warming problem through geological sequestration. Geological formations include depleted oil and gas fields, deep saline reservoirs and unminable coal seams.

CO_2 can be trapped in geological formations by three principal trapping mechanisms: (1) hydrodynamic trapping, where CO_2 can be trapped under a low-permeability caprock like gas reservoirs or aquifers, (2) solubility trapping, where CO_2 can be trapped in a dissolved phase in a liquid- like petroleum and (3) physical/mineral trapping, a relatively slower process which involves conversion of CO_2 in the form of calcium, magnesium or iron carbonates. Alternatively, in situations where CO_2 is immiscible with oil, CO_2 is injected to increase the reservoir pressure helping to push more oil towards the production well. Up to half of the injected CO_2 is stored in the immobile oil remaining in the reservoir at the end of production. The rest is collected from the production well and get re-circulated. This improves the overall economics for sequestration projects.

Gas fields have much higher primary recovery rates (80-95 per cent) than oil fields. This leaves a big void space in the reservoirs, which can be used for CO_2 storage as a supercritical gas for thousand of years. Similarly, the void space that had previously been occupied by oil and natural gases is being used for large-scale sequestration of CO_2.

A large amount of underground water filled strata (aquifers) is too salty to be used for agriculture or human consumption. These aquifers can potentially be used as long-term CO_2 reservoirs. CO_2 injected (with techniques similar to those for gas and oil fields) into these aquifers would displace brine and some of it would get partially dissolved. A part of the injected CO_2 is also reported to react with calcite and alumino-silicates to form permanent carbonates. The best example of CO_2 storage in deep saline aquifer is the Sleipner project in the North Sea that sequesters approximately 1 Mt CO_2 annually.

Storing CO_2 deep into unminable coal seams appears to be a good approach due to its value added benefit of CO_{2-}enhanced coal bed methane (ECBM) recovery. Coal beds typically contain large amounts of methane-rich gas that is adsorbed onto the surface of the coal. CO_2 adsorbs more strongly on the micropores of coal than methane (CH_4). However, the volumetric ratio of adsorbable CO_2:CH_4 depends on the type of coal. This ratio ranges from 1 for anthracite to about 10 for lignite coal. This can be exploited to lock CO_2 permanently on the micropores of coal provided the coal is never mined. Over 100,000 tons of CO_2 has been successfully injected at Allison Unit in New Mexico, USA during a ECBM project. Continuous monitoring along with exhaustive geophysical and geochemical study is, however, needed to make sure the injected CO_2 stays in ground.

Drawbacks Associated with Ocean and Geological Sequestration

Permanence of the stored carbon through a biotic sequestration method is of great critical concern. Ocean and geological storage of carbon dioxide is associated with future risk of leakage from the site of injection. Sequestered CO_2 may leak back into the atmosphere and impose future climate damages. If CO_2 migrates out of the receiving geological formation and rises to the surface, it could cause local ecological damage, primarily by displacing soil gas and affecting plant roots. Moreover, upward migration of injected CO_2 could contaminate hydrocarbon reservoirs or surface drinking water supplies. In rare cases, rapid escape of CO_2 may cause asphyxiation or toxicity risks to local animal and human populations. The eruption of CO_2 during 1986 at Lake Nyos, West of Cameroon is the most evident example, which killed more than 1700 people.

Deep-sea organisms are highly sensitive to any environmental disturbances. Increased partial pressure of CO_2 (hypercapnia) and decreased pH of seawater caused by CO_2 dissolution may affect the whole marine biodiversity. The scientific community is trying to get rid of leaky sequestration approaches. Novel concepts are being contemplated to find the most environment-friendly way to sequester CO_2. This includes the art of exploiting natural biological ways of capturing carbon and storing it in the most eco-compatible way.

BIOLOGICAL WAYS OF CARBON SEQUESTRATION

Biological systems have solutions to the most dreaded problems of all times. The photosynthetic fixation of atmospheric CO_2 in plants and trees could be of great value in maintaining a CO_2 balance in the atmosphere. Algal systems, on the other hand, being more efficient in photosynthetic capabilities are the choice of research for solving global warming problem. The biomass thus produced could be used as fuel for various heating and power purposes.

Mankind is indebted to microbes for bringing and maintaining stable oxygenic conditions on Earth. A proper understanding of microbial systems and their processes will help in stabilizing atmospheric conditions in future too. Investigations are in progress to exploit carbonic anhydrase and other carboxylating enzymes to develop a promising CO_2 mitigation strategy. Recent work on biomimetic approaches using immobilized carbonic anhydrase in bioreactors has a big hope for the safe future.

The process of carbon assimilation by photosynthesis has made forests, trees and crops as the major biological scrubbers of CO_2. Terrestrial biomes are potential CO_2 sinks. Afforestation and reforestation leads to a net increase in plant carbon stocks. A young growing forest sequesters more carbon than a matured one. Forest management can contribute to carbon sequestration by promoting forest growth and biomass accumulation. Improved cropland management (including agronomy, nutrient management, tillage/residue management and water management) has significant carbon sequestration potential. Worldwide adoption of best management practices can sequester a considerable part of the lost carbon back into croplands.

Grasslands cover about 70 per cent of the world's agricultural area. Recent studies have suggested that tropical grasslands and savannas sequester approximately 0.5 Gt of carbon annually. Grazing and burning have, however, resulted in increased soil organic carbon storage. Urban trees play a major role in sequestering CO_2. One tree in urban area is equivalent to three to five forest trees. The average sequestration rate of an urban tree of $50m^2$ crown area has been estimated to be about 11-19 kg/ year.

Plants assimilate carbon through the process of photosynthesis and return some of it to the atmosphere through respiration. After the death and decomposition of plants, carbon in the form of plant tissue is either consumed by animals or added to the soil as litter. The primary way that carbon is stored in the soil is as soil organic matter (SOM), which is a complex mixture of carbon compounds, consisting of decomposing plant and animal tissue, microbes (protozoa, nematodes, fungi, and bacteria) and carbon associated with soil minerals. Soils contain three times more carbon than the amount stored in living plants and animals. Increasing the soil organic carbon (SOC) by 0.01 per cent would nullify the annual increase in atmospheric carbon due to anthropogenic CO_2 emissions.

Microbial community structure and various microbial processes have been shown to directly affect carbon sequestration in soil agro ecosystems. A thorough understanding of microbial community structure and processes is required for enhanced carbon sequestration in agricultural soils. A balance between microbial community dynamics and formation, and degradation of microbial byproducts maintains the soil carbon content. Soil microbes also indirectly influence CO_2 cycling by improving soil aggregation, which physically protects SOM. Consequently, the microbial contribution to CO_2 sequestration is governed by the interactions between the amount of microbial biomass, microbial community structure, microbial byproducts, and soil properties such as texture, clay mineralogy, pore-size distribution and aggregate dynamics. Fungi and bacteria are responsible for most of the carbon transformations and long-term storage of carbon in soils. However, chances of persistent carbon storage are more in fungi due to their complex chemical composition and higher carbon utilization efficiency.

CARBON CONCENTRATING MECHANISMS (CCM)

Photoautotrophic organisms ranging from bacteria to higher plants have evolved with unique carbon concentrating mechanism (CCM) in response to the declining levels of CO_2 in their surrounding environment. It is proposed that ribulose-1,5-biphosphate carboxylase/oxygenase (Rubisco) co-evolved during the process. The organization of the carboxysomes in prokaryotes and of the paranoids in eukaryotes, and the presence of membrane mechanisms for inorganic carbon (C_i) transport are central to the concentrating mechanisms. There can be different types of CCM based on the biochemical mechanisms in different photoautotrophic organisms such as C_4 photosynthesis and crassulacean acid metabolism (CAM) in terrestrial higher plants, active transport of inorganic carbon (C_i) primarily in cyanobacteria and CO_2 concentration following acidification in a compartment adjacent to Rubisco found in some eukaryotic algae.

Higher terrestrial plants having crassulacean acid metabolism (CAM) primarily capture CO_2 through PEP carboxylase located in the cytosol of their mesophyll cells. PEP carboxylase uses bicarbonate as its primary substrate for fixation of CO_2 into oxaloacetate, thus CO_2 entering from the external environment must be hydrated rapidly by a carbonic anhydrase (CA) and converted to bicarbonate. C_4 carboxylic acids such as malate or aspartate formed in the mesophyll cell cytosol serve as the intermediate CO_2 pool.

CCM found in eukaryotic algae relies on the pH gradient set up across the chloroplast thylakoid membrane in the light. Light-driven photosynthetic electron transport sets up a pH around 8.0 in chloroplast stroma and a pH between 4 to 5 inside the thylakoid lumen. Under these conditions, bicarbonate is the predominant species of C_i in the chloroplast stroma, while CO_2 is the most abundant form of C_i in the thylakoid lumen. Bicarbonate transporters on thylakoid membrane are proposed to help bicarbonate transport inside thylakoid lumen where it is converted into CO_2 with the help of carbonic anhydrase. Microalgal mass cultures can use CO_2 from power plant flue gases for the production of biomass. The algal biomass thus produced can directly be used as health food for human consumption, as animal feed or in aquaculture, for biodiesel production or as fertilizer in agriculture. A fast growing marine green alga *Cholococcum littorale* is reported to tolerate high concentrations of CO_2. Waste water containing phosphate (46 g m^{-3}) from a steel plant has been to raise cultures of the photosynthetic microalgae *Chlorella vulgaris*. Flue gas containing 15 per cent CO_2 was supplemented further to get a CO_2 fixation rate of 26 g CO_2 $m^{-3}h^{-1}$. Research is in progress on the development of a novel photo bioreactors for enhanced CO_2 fixation and $CaCO_3$ formation. CO_2 fixation rate was increased from 80 to 260 mg $l^{-1}h^{-1}$ by using *Chlorella vulgaris* in a newly developed membrane-photo bioreactor. A novel multi-disciplinary process has recently been proposed using algal biomass in a photo bioreactor to produce H_2 besides sequestering CO_2. Enhanced growth rate of marine macro algae such as *Gracilaria* sp. and *G. chilensis* has been observed by increasing CO_2 concentration from 650 ppm to 1250 ppm. The macroalgal culture can make important contribution to both biomass production for chemicals and fuel besides CO_2 remediation.

Photosynthesis is much more efficient in microalgae than in terrestrial C_3 and C_4 plants. This high efficiency is again due to the presence of both intracellular and extracellular carbonic anhydrases and the CO_2 concentrating mechanism. The present focus is on exploiting the ability of microalgae to convert solar energy and CO_2 into O_2 and carbohydrates. Considerable efforts have been made for CO_2 fixation along with valuable material production by mass cultivation of algal cultures.

Non photosynthetic CO_2 fixation occurs widely in nature by the methanogenic archaebacteria. These are obligate anaerobes that grow in freshwater and marine sediments, peats, swamps and wetlands, rice paddies, landfills, sewage sludge, manure piles and the gut of animals. Methanogens are responsible for more than half of the methane released to the atmosphere. These methanogenic bacteria grow optimally at temperatures between 20°C and 95°C. Carbon monoxide dehydrogenises and/or acetyl-COA synthesis aid them to use carbon monoxide or carbon dioxide along with hydrogen as their sole energy source.

Waste gases from blast furnaces containing oxides of carbon were used for converting them into higher Btu (more calorific value) methane using thermophilic methanogens. A column bioreactor operated at 55 °C and pH 7.4 was used for the process. A mixture of three culture of bacteria, *viz. Rhodospirillum rubrum, Methanobacterium formidium* and *Methanosarcina barkeri* was used for complete bioconversion of oxides of carbon to methane. Acetogenesis, on the other hand, is involved in the recycling of 10 to 20 per cent of the carbon on earth.

CARBON SEQUESTRATION USING HETEROTROPHIC BACTERIA

The concept of CO_2 fixation in certain representatives of heterotrophic bacteria was first proposed by Wood and Werkman in 1941. While working on propionic acid bacteria, they proposed that CO_2 and pyruvate combine to form oxaloacetate. The same pathway can be exploited now for capturing carbon using heterotrophic bacteria. Carbonic anhydrases play a critical role in concentrating CO_2 inside the cell. The capability of carbonic anhydrases to convert CO_2 in bicarbonate may be utilized by carboxylases such as phosphoenolpyruvate (PEP) carboxylase and pyruvate carboxylase, to form oxaloacetate. Such anapleurotic pathway exists in organisms to compensate for the loss of oxaloacetate siphoned off for the synthesis of amino acids of aspartate family.

Heterotrophic bacteria having maximal carbonic anhydrase and phosphoenol-pyruvate carboxylase and/or pyruvate carboxylase titers may be raised in fermentors, and these can be flushed with flue gases with CO_2 to produce useful metabolites such as oxaloacetate and amino acids. Extensive research has been done on the production of glutamic acid and lysine by *Corynebacterium glutamicum*. The presence of both phosphoenolpyruvate carboxylase and pyruvate carboxylase and the PEP–pyruvate–oxaloacetate node makes this bacterium suitable for fixing carbon in the form of amino acids. An increased bicarbonate supply by the action of carbonic anhydrases (or elevated CO_2 conditions) to phosphoenolpyruvate and pyruvate carboxylases may enhance their activity, thereby making the conditions favorable for enhanced lysine production. Work is in progress in our laboratory at the University of Delhi South Campus to understand the effect of different levels of carbon dioxide on carbonic anhydrase, phosphoenolpyruvate carboxylase and pyruvate carboxylase titres, and hence, their overall effect on lysine production. Dual benefit of carbon sequestration along with useful product formation makes this approach very attractive. Microbes, being widespread in nature, play a major role in chemical cycles that influence atmosphere-hydrosphere composition and are extensively involved in the production and accumulation of various sediments deep inside the oceans. Bacteria are the key organisms in the formation of microbial carbonates.

Mineral carbonation has emerged as a new carbon capture and storage technology in the past few years. The idea of applying carbonation reactions for CO_2 storage was proposed by Seifritz in 1990. Carbonic anhydrases are the fastest enzymes known for their capability and efficiency for converting carbon dioxide into bicarbonates. Gillian M. Bond of New Mexico Tech, USA initiated work on this enzyme for mineral CO_2 sequestration in 2001.

Biomimetic approach involves identification of a biological process or structure and its application to solve a non biological problem. It has emerged as an environment friendly process, which can be operated at near ambient temperature and pressure with no costly CO_2 concentration and compression steps. The process of carbon dioxide fixation can be carried out successfully with a stream of carbon dioxide (from flue gases) in a bioreactor. Various methods for carbonic anhydrase immobilization are being attempted for the development of an efficient biodegradable matrix that can ensure maximal activity along with its long-term use in bioreactors for sequestration purposes. Carbonic anhydrase was recently immobilized in chitosan-coated alginate beads. A novel trickling spray reactor employing immobilized carbonic anhydrase has been developed that enables concentration of CO_2 from the emission stream. Carbonic anhydrase is one of the fastest enzymes, which make mass transfer from the gas phase to aqueous phase. This biocatalytic fixation of carbon could be the answer to tackle atmospheric pollution.

Most of the work on biomimetic sequestration uses carbonic anhydrase from animal sources. However, there is a need for a thermophilic carbonic anhydrase that sustains high pressure if we really want to use brine as the most favorable cationic source for mineral carbonation under in the deep-sea environment. A gene encoding putative β-type carbonic anhydrase in the methanoarchaeon *Methanobacterium thermoautotrophicum* has been expressed in *E. coli* and found to encode a thermostable (up to 75°C) carbonic anhydrase. Its activity at different hydrostatic pressures needs to be studied for its biomimetic applications in carbon sequestration.

The biomimetic approach has now also been applied in relation to geological sequestration. A 'closed-loop' fossil-fuel carbon cycle has been proposed to be developed, in which microbial consortium (comprising of methanogens) could be used to convert CO_2 to methane at a commercially useful rate. This can be used either in a geological setting (following injection of CO_2 into depleted oil and gas well, saline aquifer, etc.) or above ground in rapid-contact reactors.

CONCLUSIONS

Several novel concepts and techniques are being attempted for a safe and permanent capture of CO_2. Routine abiotic methods although appear promising at *prima facie* but costly concentration and transportation steps along with future leakage risks have led to focus on new biotic methods. Evolution has equipped plants and various domains of microbial life with different mechanisms for carbon fixation. The present need is to exploit these biological mechanisms along with existing biochemical engineering techniques for long-term CO_2 sequestration. Exhaustive study needs to be done on various metabolic pathways that employ carboxylases. Behavior of enzymes like carbonic anhydrase and Rubisco with gases other than CO_2 in flue gas must be understood. Despite finite sink capacity, biological approaches provide a natural and cost-effective method of carbon sequestration. Biotic and abiotic approaches have their own merits and demerits, and they are complementary to each other and have the potential to mitigate the risks of climate change. The World Environment Day slogan for 2008 'Kick the Habit! Towards a Low Carbon Economy'– has become the defining issue of the present era.

FURTHER READING

[1] Department of Energy (DOE), 1993, A research needs assessment for the capture, utilization and disposal of carbon dioxide from fossil fuel fired power plants. DOE/ER-30194, Washington, D.C., USA.

[2] Halmann, M.M. and M. Steinberg, 1999, Greenhouse gas carbon dioxide mitigation: science and technology, Lewis publishers, Boca Raton, Florida. pp. 1-3.

[3] Held, I. M. and B. J. Soden, 2006, Robust responses of the hydrological cycle to global warming. Journal of Climate, 19: 5686-5699.

[4] Holloway, S., 2005, Underground sequestration of carbon dioxide– a viable greenhouse gas mitigation option. Energy, 30: 2318-2333.

[5] House, K.Z., D.P. Schrag, C.F. Harvey and K.S. Lackner, 2006, Permanent carbon dioxide storage in deep-sea sediments, Proc. Natl. Acad. Sci. USA, 33: 12291-12295.

[6] O'Neill, B. C. and M. Oppenheimer, 2002, Climate change– dangerous climate impacts and the Kyoto protocol, Science, 296 (5575): 1971-1972.

[7] Gilfillan, S.M.V. *et al.*, 2009, Solubility trapping in formation water as dominant CO_2 sink in natural gas fields, Nature, 458: 614–618.

[8] Goel, M., 2007, Carbon capture and storage technology for sustainable energy use, Curr. Sci., 92: 1201-1202.

[9] Jadhav, P.D., S.S. Rayalu, R.B. Biniwale and S. Devotta, 2007, CO_2 emission and its mitigation by adsorption on zeolite and activated carbon, Curr. Sci., 92: 724–726.

[10] Bond, G.M. *et al.*, 2001, Development of integrated system for biomimetic CO_2 sequestration using the enzyme carbonic anhydrase, Energy Fuels, 15: 309–316.

Chapter 11

Opportunities for Extraction and Utilization of Coal Mine Methane and Enhanced Coal Bed Methane Recovery in India

Ajay Kumar Singh

Central Institute of Mining and Fuel Research, Barwa Road, Dhanbad – 826 015

SUMMARY

Emission of methane in coal mines has been a source of safety hazard throughout the globe. Sufficient quantity of air is sent to the mines through ventilating fan to dilute the methane concentration significantly below the statutory level. Indian mine regulation permits maximum 0.75 per cent of methane concentration in the return air and 1.25 per cent at other places. In degree II and III gassy mines, where the gas content of coal is higher, emission of methane in the workings increases with increased coal production. Eventually, the ventilation system of the mine may become inadequate to maintain the methane concentration within the permissible limit. Methane is, therefore, a limiting factor for concentrated production of coal, particularly in gassy mines. On the other hand, methane released in a mine is vented out to the atmosphere, causing accumulation of this potent greenhouse gas (GHG) in the environment.

Coal Mine Methane (CMM) and Ventilated Air Methane (VMM) thus have vast potential as future energy sources. Degasification methods can be deployed to harness CMM in the Indian coal fields. Current efforts in exploration of CMM and AMM at the Central Institute of Mining and Fuel Research (CIMFR) and methodology for CMM resource evaluation·being developed for Indian Geo-

Mining conditions described. There are also various methods for extraction of abandoned coal mine methane (AMM) through boreholes and vents. This paper discusses about CBM development in India, various forms of CBM, and the status of CBM and Enhanced Coal Bed Methane (ECBM) activities in the country.

INTRODUCTION

Emission of methane in the underground coal mines has been a source of safety hazard throughout the globe. Many of the disasters have occurred in the past due to explosion of fire damp. Among the disasters in coal mines, 42 per cent of them were caused by methane explosions, leading to 41 per cent of the total casualties. It is established that mine ventilation air with methane concentration in the range of 5 to 15 per cent by volume, when exposed to a source of ignition causes explosion. Eventually, the ventilation system of the mine may become inadequate to maintain the methane concentration within the permissible limit. Methane is, therefore, a limiting factor for concentrated production of coal, particularly in gassy mines.

Of late, it has been established that coal bed methane (CBM) can be recovered at commercial scale and used as a clean source of fuel. The CBM industry started in the USA in 1980s, and now there are more than 12,000 wells producing 9 per cent of the total USA natural gas production. After success in the USA and later in Australia, CBM has attracted worldwide attention as an alternate source of energy. Exploration and production activities have started in other coal producing countries including India, China, Russia, Poland and other European countries. In India, exploration activity was started in the mid 1990s and since then significant progress has been achieved in developing the CBM industry. This paper discusses about CBM developments in India, various forms of CBM, and the status of CBM and Enhanced Coal Bed Methane (ECBM) activities in the country.

METHANE IN COAL

Large amounts of gases are generated by chemical and biological processes during coalification of plant materials. The product gas mostly contains methane (80-95 per cent) and is known as coal bed methane, although other gases such as CO_2, N_2, He, and other hydrocarbons may be present in varying concentrations [1,2]. The quantity of methane generated depends on the degree of coalification and other geological factors. The amount of methane generated during the conversion of lignite to anthracite may be as large as 200 m^3/t and greatly exceeds the capacity of the coal to hold the gas. Most of the gas migrates to the atmosphere through the surrounding strata. Only a portion is retained in the coal and this is known as the gas content of coal. The amount of gas contained in coal depends on the geo-environmental parameters like pressure, temperature and moisture content of coal, chemical parameters like nature, rank and composition, and reservoir properties like permeability, porosity, degree of fracturing of coal and adjacent rocks.

Coal is a microporous solid possessing a large internal surface area, typically 20 to 200 m^2/g [3]. Gas stays in coal in three different phases, *viz.* adsorbed gas, free gas and dissolved gas. Physical sorption is the most dominant mechanism accounting for more than 90 per cent of the gas stored in coal. The sorbed gas stays in the micropores in a compressed state with liquid-like density [4].

DIFFERENT CATEGORIES OF CBM

From extraction technology and utilisation perspective, coal bed methane is classified into the following four categories.

1. Virgin Coal Bed Methane (VCBM) or Pre-mining degasification: This terminology is used when the CBM is recovered from virgin coal seams prior to the mining activity.
2. Coal Mine Methane (CMM) or Degasification of working seams: When methane is recovered simultaneously during mining of coal with an objective of reducing the methane concentration in the mine workings and for its utilisation, it is known as the coal mine methane. Various mine degasification techniques are followed to recover CMM.
3. Ventilation Air Methane (VAM): Methane present in the ventilation air is known as the ventilation air methane. Although, the concentration of methane in the ventilation air is generally very low, new oxidation technologies have come up which burn the VAM and produce useful energy from it.
4. Abandoned Mine Methane (AMM): Methane that continues to be emitted from the left out coal and the adjoining strata and is accumulated in the mine void even after the abandonment of an underground mine, is known as abandoned mine methane. In some of the gassy mines, the AMM may be of significant volume and quality, which can be used for economic purposes such as power generation.

While exploiting virgin coal bed methane is the activity of natural gas industry, capturing CMM, VAM and AMM is practiced by the mining industry. All the above four different categories of methane from coal and coal mines are discussed separately in the following sections.

VIRGIN COAL BED METHANE

The retention mechanism of methane in coal being physical adsorption is a revertible process. Therefore, the most common practice for recovering coal bed methane in a virgin block is to depressurize the coal bed, usually by pumping water out of the reservoir. Vertical wells are drilled across the coal seam typically in a five spot pattern and water is pumped out continuously. Sometimes well is stimulated by hydraulic fracturing to improve the permeability of the reservoir. This technique is known as the primary method of recovery. In the beginning, there is only production of water. Afterwards, when the reservoir pressure comes down to desorption pressure, production of methane starts along with that of water. Subsequently, water production decreases and production of methane continues for a longer time, typically 20 to 25 years. Generally 20 to 60 per cent of the gas-in-place can be recovered by the primary method [5-7]. The remaining gas, which could not be recovered by primary method may, in fact, be recovered by injecting a second gas, such as CO_2 or N_2 for pressure maintenance. Enhancement of coal bed gas recovery by such injection is known as the enhanced coal bed methane (ECBM) recovery technique. Enhancement using nitrogen is based on the principle of inert gas stripping, where the injected N_2 reduces the partial pressure of methane and induces desorption, thus resulting in greater recovery [7]. The second ECBM technique utilizes injection of a higher adsorbing gas, such as CO_2, which is preferentially sorbed. The injected CO_2 displaces methane from the site of adsorption in coal, resulting increased production of methane, while CO_2 gets adsorbed in the coal micropores and is sequestered permanently [4,8]. Recovery using this technique may go up to 70-90 percent of the gas-in-place [4,9].

Status of CBM Activity in India

In India, the coal bed methane resource exploration is regulated by the Ministry of Petroleum and Natural Gas, Government of India. Blocks have been allotted to the public and private sector companies by three rounds of global bidding, CBM I, CBM II, and CBM III. The awarded blocks are shown in Figure 1.

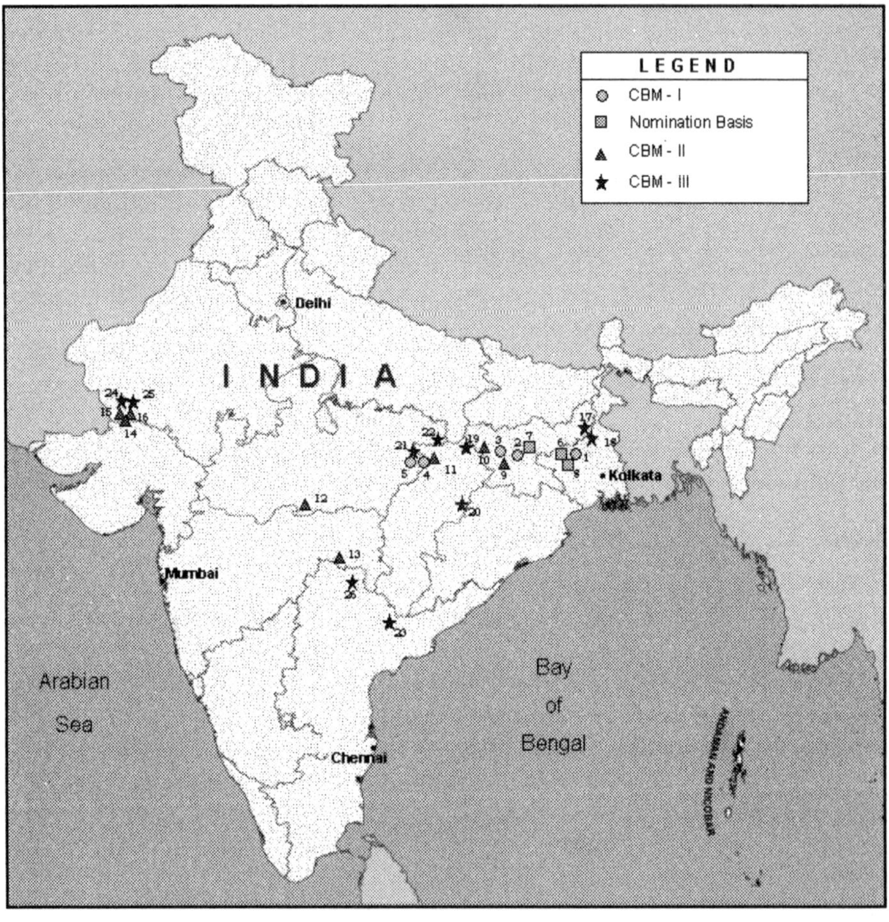

Figure 1: Coal Bed Methane Blocks Awarded by MoPNG
(*Source*: DGH, 2005-06)

So far 26 CBM blocks spread over an area of 13,600 sq. km have been opened up for exploration by these three CBM rounds including three blocks awarded on nomination basis. Till now 70 boreholes have been drilled and 40 pilot wells have been started involving an investment of 170 crores. CIMFR has played a very important role in ascertaining the resource base of CBM in the country by undertaking the gas desorption, adsorption and other relevant studies in different coalfields throughout the country. Based on the investigations carried out by CIMFR, the coal and lignite fields in India may be classified into four different categories shown below in Table 1.

Table 1: Categorization of Indian Coals for CBM Potential

Category	*Coal Type*	*Coalfields*
Category-I	Gondwana coals ranking high volatile bituminous A and above.	Jharia, East and West Bokaro, Raniganj and North Karanpura Coalfields
Category-II	Gondwana coals ranking high volatile bituminous A and below.	South Karanpura, Raniganj, Pench- Kanhan and Sohagpur Coalfields.
Category-III	Low rank Gondwana coals	Talchir, Ib valley, Pranhita-Godavari Valley, Wardha Valley Coalfields.
Category-IV	Tertiary Coals/Lignite	Assam-Arakan, Himalayan Foothills, Cambay, Barmer-Sanchor, Bikaner- Nagaur and Cauvery basins.

Category I

These are Gondwana coals ranking high volatile bituminous A coals and above available in the Jharia, Bokaro, Raniganj and North Karanpura coalfields. These coals have the best CBM potential in India. Estimated resource in these basins is 350–400 BCM and a producible reserve of 85–100 BCM.

Category II

Gondwana coals ranking high volatile bituminous A and below in the South Karanpura, Raniganj, Pench-Kanhan and Sohagpur coalfields fall in the category II.

Category III

Low rank Gondwana coals in Talcher, Ib valley, Korba, Pranhita-Godavari and Wardha valley coalfields may be classified as category III. However, coals in these basins occur at shallower depth, have extensive thickness and may provide suitable hydro dynamic conditions for recovery.

Category IV

Category IV basins constitute tertiary coals and lignite and have lowest CBM resource base. However, better permeability and reservoir properties make these basins suitable for production.

The total CBM resource in the country is estimated to be 1374 TCM [10]. The Great Eastern Energy Corporation Limited (GEECL) an energy company based at Kolkata is at the threshold of commercial production from Raniganj coalfields near

Burnpur. Oil and Natural Gas Corporation (ONGC) production has been doing production testing in four boreholes at Parbatpur in the Jharia Coalfield. The production data are very encouraging and commercial production is very likely to start at Parbatpur from the year 2008. Similarly Reliance India Limited is also undertaking production testing in a few boreholes at Sohagpur area in Chhatishgarh. Essar Oil Limited is about to complete the exploration in the Raniganj Coalfield.

COAL MINE METHANE

Recovery of methane from active mine is known as mine degasification. Mine degasification involves removing methane from the coal seam and surrounding strata prior and during the mining of coal using boreholes or wells. Mine degasification techniques have been applied in several countries like USA, Australia, Germany and others in order to reduce the production of methane in general body of mine air of underground coal mines and goaf area. These techniques have been successful in several cases, in keeping the methane concentration within the stipulated limit and also reducing the ventilation cost. Mining degasification methods are generally classified into four categories as listed below. The degasification methods include, but not limited to:

- ☆ Vertical wells
- ☆ Goaf wells
- ☆ Horizontal boreholes
- ☆ Cross-measure boreholes

Vertical Wells

Vertical wells are drilled through coal seams to pre-drain methane prior to mining. The drainage operation is started more than 2 years ahead of mining. These wells are drilled from surface on a grid pattern to intersect coal seams to be mined in the future. The vertical wells are generally coupled with hydraulic fracturing of the coal seam to improve the recovery of methane. Sometimes open-hole cavity completions are also done to enhance the flow of methane from the seam. The technique has yielded gas flows up to 7,000 m^3/day from a well in certain coal seams [11]. In case of hydraulic fracturing, both vertical and horizontal fractures are created in the coal seam. The amount of gas recovered by vertical wells depends on a host of factors like, the gas content of coal and surrounding strata, permeability, amount of negative head applied, drainage time allowed and other geologic and operation parameters. Vertical wells can recover 50 to 90 per cent of the gas content of the coal.

Goaf Wells

Drilling above the goaf of mines is done in order to recover methane produced in the goaf. These wells are known as goaf wells and are drilled to a point 10 to 50 ft. above the target seam prior to mining, but are operated only after mining fractures the strata around the well bore. The flow rate is controlled by the natural head created by methane and can be stimulated by a suction fan on the surface. Goaf wells can recover 30 per cent to 70 per cent of methane present in the seam depending on geologic conditions and the number of goaf wells within the panel. Goaf wells have several disadvantages compared to vertical wells. One is that the gas quality (content of methane in the gas) varies over the well life. Also the production life of a goaf well is typically short and lastly, it drains gas only from the overlying strata. The goaf wells have been widely used in longwall mines in USA.

Horizontal Boreholes

Horizontal boreholes are drilled into the coal seam from development entries in the mine shortly before mining the coal. They are typically 300 to 900m in length [11]. The life span of a typical horizontal borehole varies between one to two years. Since the time available for drainage of gas is short, the recovery efficiency is also lesser, in the range of 10 to 20 per cent. The gas recovered by horizontal borehole system is typically of high quality.

Cross-Measure Boreholes

Cross-measure boreholes are drilled at an angle to the overlying or underlying strata from the development entries with a purpose of pre-draining the methane from these and exhausting gas from the goaf area. This system is extensively followed in Eastern Europe. The angle at which the holes are drilled is a function of the height and width of the geologic zone to be drained and the location of the entries from which the holes are to be drilled.

CMM Activity in India

There is good potential of recovery of CMM in gassy mines of Jharia and Raniganj coalfields, particularly in the degree III gassy mines. During the late seventies, studies were undertaken to assess the efficacy of the various degasification techniques for some of the gassiest underground coal mines of India. Feasibility reports were also prepared for such mines. The mines for which such studies were undertaken are Amlabad and Moonidih in the Jharia coalfield and Ghusick in the Raniganj coalfield. All the different degasification techniques were applied. Presently United Nations Development Programme is funding a CMM recovery capacity building project at Moonidih and Sudmadih mines of Bharat Coking Coal Ltd. (BCCL). Central Mine Planning and Design Institute (CMPDI) and BCCL are the implementing agencies with CIMFR as a partner. Horizontal drilling of boreholes are planned for recovery of CMM from longwall panels at Moonidih and Sudamdih mines. Through the Methane to Markets (M2M) partnership, the US Environmental Agency (USEPA) has funded a project on feasibility study for recovery and utilization of coal mine methane in Jharia, Bokaro and Raniganj Coalfields. The study is being conducted by CIMFR. Further, CIMFR along with Molopo Australia is also planning a study to evaluate the CMM potential in some of the gassy mines in Jharia and Raniganj Coalfields. CMM recovery will actually be demonstrated at a potential site in India.

VENTILATION AIR METHANE

Air coming out of the ventilation air shaft contains methane, though the concentration is very low. Combustion of VAM to generate useful energy has been thought for long. However, utilization of VAM has been technically challenging due to several reasons. Ventilation air typically contains low and variable methane concentration and large quantity of the ventilation air make it difficult to handle and process it into useable forms of energy. Of late a few technologies have been developed to utilise VAM for generation of electricity. The technologies for utilisation of VAM are classified into two broad categories.

(*i*) Ancillary use technology, and

(*ii*) Principal use technology

In the ancillary use technology, ventilation air instead of the ambient air is used to provide the oxygen needed for the combustion of the primary fuel in Internal Combustion (IC) engines, turbines, boilers, furnaces or any other combustion unit. Ancillary technology has been successfully used in the Appin colliery in Australia. However, this system may also employ more concentrated fuels such as goaf gas to enhance the utility or profitability of the VAM utilization project. Two major principal use technologies reported in the literature are: (*i*) thermal flow-reversal reactor (TFRR), and (*ii*) catalytic flow-reversal reactor (CFRR). The TFRR system utilizes the thermal oxidation principle to generate heat and then the heat of oxidation is converted to electrical power. VOCSIDIZER developed by the MEGTEC Inc. is the most common TFRR. The VOCSIDIZER operates above 1000°C, the auto ignition temperature of methane. The VOCSIDIZER is a simple apparatus that consists of a large bed of silica gravel or ceramic heat exchange medium with a set of electric heating elements in the centre. The process employs the principle of regenerative heat exchange between the ventilation air and a bed of silica gel to store and transfer heat efficiently in the reaction zone. CFRR is an improvement over the TFRR in that, it could process mine ventilation air at lower temperatures. In this reactor, a suitable catalyst is used to reduce the auto ignition temperature of methane by several hundred degrees Celsius. A Canadian consortium has demonstrated the CFRR at industrial scale.

VAM Activity in India

India has about 17 degree III mines and there is a scope of utilization of VAM in these mines. CIMFR along with Southern Illinois University Carbondale (SIUC) is undertaking a study to evaluate the resource base and potential of utilization of VAM at some of the gassy mines in the Jharia and Raniganj Coalfields. Moonidih and Sudamdih mines of BCCL have been chosen in the first phase to undertake the study.

Technology exists for utilisation of both mine ventilation air and waste coal. Waste coal of different combination depending on availability of carbon and calorific value are readily available as supplementary fuel at all mine sites. A kiln coupled to an externally fired steam turbine has been developed by CSIRO, Australia for utilisation of ventilation air methane with a combination of coal middlings. Using this method, gas and coal are combusted in a rotating kiln, which is capable of burning coal of high ash content and low concentration methane in ventilation air. The advantage of this approach is that if methane levels fluctuate, more waste coal can be added to the system to even out the energy supply from the ventilation air. The system is extremely flexible and allows a mine to adapt the system to its own requirements. In this system, the combustion chambers of the gas turbine are replaced by a heat exchanger that recovers the heat of combustion. Mines with a lot of gas could operate almost as efficiently as modern combined cycle power plant, despite deriving almost two-third of the energy from high ash waste coal. Opportunities for utilisation of Hybrid Coal and Gas Turbine System (HCGTS) have been evaluated and a joint project proposal of BCCL, Tata Projects and CIMFR for implementation of this technology at Moonidih is envisaged.

ABANDONED MINE METHANE

AMM is generally the emitted gas from left out coal pillars and surrounding strata of abandoned and idle mines. As active mining stops, the mine's gas production decreases, but does not stop completely. Following an initial decline, abandoned mines can liberate methane at a near-steady rate over an extended period of time. The abandoned mines emit gas to the atmosphere through mine openings or factures that connect the mine void to the surface and through leaking or vented seals that are placed over ventilation shafts and other openings. Some of the abandoned mines may flood as a result of intrusion of groundwater or surface water into the void. Flooded mines typically produce gas for only a few years. Formation of AMM is affected by the method in which the mine is sealed at the time of closure or abandonment, the gas evolution rate from the interior strata, leakage from adjacent mine workings, the degree of flooding that occurs in the mine after closure, atmospheric pressure changes, diurnal changes, filling of mine shafts with solid material, and other factors. AMM can be extracted by suction of gas through: (*i*) boreholes drilled from the surface; (*ii*) pipes that were emplaced through isolating seals within a mine's districts, sections or roadways; or (*iii*) vents installed in mine shaft seals.

AMM Status in India

There has not been any effort to identify the abandoned mines and quantify the AMM resource in India. It is thus, imperative to undertake a study to evaluate the AMM resource potential of the country before any utilization potential is planned. CIMFR will undertake the study through a project funded by the Ministry of Environment and Forests, Government of India.

ENHANCED COAL BED METHANE RECOVERY

Deep and unmineable coal seams or Coal Bed Methane (CBM) reservoirs are another potential CO_2 storage site. Primary CBM recovery technique leads to 20-60 per cent recovery of the gas-in-place [6]. Some of the remaining CBM may be further recovered by injecting CO_2. Coal has a preferential sorption affinity for CO_2 over methane with an adsorption ratio ranging between 2:1 to 8:1 [8,12]. The injected CO_2 molecules displace the adsorbed methane molecules which desorb from the coal matrix into the cleats and flow to the production wells. The disposal of CO_2 in the coal beds is expected to increase drive pressure and the CBM recovery rate [13]. CO_2 injection can achieve about 70-90 per cent recovery of gas-in-place [14]. Simultaneously, the injected CO_2 diffuses through the pore structure of coal and is physically adsorbed to it and is thus, stored securely.

CO_2 sequestration in coal seams has the potential to generate cash flow through additional CBM production. The total worldwide potential for CBM is estimated at around two trillion standard cubic meter (scm) with about 7.1 billion tons of associated CO_2 storage potential. The enhanced recovery of CBM may make the storage operation very cost-effective or even cost free. Although, the sequestration of CO_2 in coal seams is an attractive option, the physical coal-CO_2 interactions are largely unknown. Recent studies have shown that injection of CO_2 in coal beds may result in decreased permeability of the cleat system surrounding the injection well area because of swelling of coal [15]. Such swelling may partially block the cleat system and negatively affect the flow parameters.

ECBM Status in India

Nearly 99.7 per cent bituminous to sub bituminous coal of India is available in Lower Gondwana in the eastern region of India: in the States of West Bengal, Jharkhand, Madhya Pradesh, Chhattisgarh, Orissa, Andhra Pradesh and Maharashtra. There are more than 65 known basins in Lower Gondwana sediments spread over nearly 64,000 sq.km. Excluding unproductive part and small detached outliers the potential coal bearing area is about 14,500 sq.km.

The potential sites for CO_2 storage in coal beds of Indian Territory have been identified with due consideration of accepted exploration norms, depth wise resource distribution quality wise abundance and mining status of coal. The following areas in different coalfields appeared to have fair chance of CO_2 storage (Table 2).

Table 2: Identified Candidates for CO_2 Storage in Indian Territory

Category of Coal Beds	*Grade of Coal*	*Candidates/Basins*
Unmineable coal beds	Power Grade coal	Singrauli, Mand Raigarh, Talcher, Godavari
Grey Areas	Coking coal	Jharia, East Bokaro, Sohagpur, S Karanpura
	Superior non coking coal	Raniganj, South Karanpura
	Power grade coal	Talcher
Concealed Coalfields	Tertiary age coal	Cambay basin, Barmer Sanchor basin
	Power grade coal	West Bengal Gangetic Plain, Birbhum, Domra Panagarh, Wardha Valley Extension, Kamptee basin Extension.

CONCLUSIONS

Recovery of Coal Mine Methane (CMM) and Enhanced Coal Bed Methane (ECBM) recovery offers the possibility of continuing use the low cost fossil energy sources while developing efficient energy technology and shifting to cleaner energy sources. Option of CMM exploitation and CO_2 storage in deep unmineable coal seams look attractive at the moment for various reasons. These include economic benefits, however, their technical feasibility is yet to be proven.

ACKNOWLEDGEMENT

The author gratefully acknowledges the US Environmental Protection Agency (USEPA) for providing financial support. Thanks are also due to Dr. Amalendu Sinha, Director, CIMFR, Dhanbad for his encouragement and fruitful discussion.

REFERENCES

[1] Greaves, K.H., L.B. Owen and J.D. McLennan, 1993, Multi-component gas adsorption–desorption behavior of coal, Paper 9353, International Coal Bed Methane Symposium, Tuscaloosa, AL.

[2] Kim, A.G.,1977, Estimating methane content of bituminous coal beds from adsorption data, U.S. Bureau of Mines Report of Investigation, RI 8245.

[3] Yee, D., J.P. Seidle and W.B. Hanson, 1993, Gas sorption on coal and measurement of gas content, In B.E. Law, and D.D. Rice (Eds.). Hydrocarbons from coal. AAPG Studies in Geology: Vol. 38 (Chap. 9, pp. 203–218), American Association of Petroleum Geologists, Tulsa.

[4] Gunter, W.D., T. Gentzis, B.A. Rottenfusser and R. J. H.Richardson, 1997, Deep coal bed methane in Alberta, Canada: A fuel resource with the potential of zero greenhouse gas emissions, Energy Conversion and Management, 385, 217–222.

[5] Voormeij, D.A., and G. J. Simandl, 2003, Geological and mineral CO_2 sequestration options: A technical review. British Columbia Geological Survey, Geological Fieldwork 2002, 265-276.

[6] Gentzis, T., 2000, Subsurface Sequestration of Carbon Dioxide- An Overview from an Alberta (Canada) Perspective. International Journal of Coal Geology, Vol. 43, p. 287-305.

[7] Puri, R. and D. Yee, 1990, Enhanced coal bed methane recovery. SPE 20732, SPE 65th Annual Technical Conference and Exhibition, New Orleans, LA, 193-202.

[8] Arri, L.E., D. Yee, W.D. Morgan and M.W. Jeansonne, 1992, Modeling Coal Bed Methane Production with Binary Gas Sorption, Society of Petroleum Engineers Paper, No 24363, p. 459-472.

[9] Wong, S., W.D. Gunter and M.J. Mavor, 2000, Economics of CO_2 sequestration in coal bed methane reservoirs. SPE 59785, SPE/CERI Gas Technology Symposium, Calgary, Canada, 631-638.

[10] Gautam, N.N., 2006, Coal Bed Methane–Present and Prospects, International Coal Congress and Expo 2006, 11-13 Dec, New Delhi, India.

[11] Thakur, P.C., H.G. Little and W.G. Karis 1996, Global Coal bed Methane Recovery and Use, Energy Conversion and Management, Vol. 37, Nos 6-8, pp. 789-794.

[12] Krooss, B.M., F. van Bergen, Y. Gensterblum, N. Siemons, H.J.M. Pagnier and P. David, 2002, High-Pressure Methane and Carbon Dioxide Adsorption on Dry and Moisture Equilibrated Pennsylvanian Coals, International Journal of Coal Geology, Vol. 51, p. 69-92.

[13] Hitchon, B., W.D. Gunter, T. Gentzis and R.T. Bailey, 1999, Sedimentary Basins and Greenhouse Gases: A Serendipitous Association, Energy Conversion and Management, Vol. 40, No. 8, p. 825-843.

[14] Reeves S., 2004, The Coal-sequestration Project: Key Results from Field, Laboratory, and Modeling Studies, International Conference on Greenhouse Gas Control Technology, Vancouver, Canada.

[15] Reeves, S., 2002, Coal-Seq Project Update: Field Studies of ECBM Recovery/CO_2 Sequestration in Coal Seams, Proceedings of the 6th International Conference on Greenhouse Gas Control Technologies, Elsvier, London UK, (J1-2).

Chapter 12

Coal Bed Methane: Prospects and Challenges

V.A. Mendhe

Central Institute of Mining and Fuel Research, Barwa Road, Dhanbad – 826 015

SUMMARY

India is rich in coal and is the third largest coal producer in the world. Coal continues to be the dominating energy source and meets nearly 58 per cent of total requirement of commercial energy. India has huge Gondwana (mainly Permian–99.5 per cent) and Tertiary (Eocene and Oligocene) coal deposits distributed in several basins located in peninsular and extra-peninsular regions of about 257 billion tons. The prospect for Coal Bed Methane (CBM) is mainly related to the coal resources of the country. CBM an unconventional source of natural gas is now considered as an alternative source for augmenting the country's energy resources. The environmental, technical and economic advantages of CBM have made it a global fuel of choice. Indian scenario of coal bed methane (CBM) in different coal seams is described in this paper. The exploration carried out by various operators in Raniganj, Jharia, Bokaro, North Karanpura and Sohagpur coal bearing provinces reveals that Barakar/Raniganj coal reserves are excellent CBM players, which can be harnessed commercially with the induction of appropriate technology. In the Jharia Coalfield, the gas content is estimated to be between 7.3 and 23.8 m^3 per tons of coal within the depth range of 150 to 1200 m. Analysis indicates every 100-m increase in depth is associated with a 1.3 m^3 increase of methane content. The scientific research has been undertaken in Moonidih and Jharia coals and work has also begun to understand coal morphology for CO_2 sequestration. This area coal of research is still in infancy worldwide and has a long-term perspective.

INTRODUCTION

Energy resources and its best utilization decide the industrial growth and prosperity of a country. Coal is basic source of energy for the industrial development in the world. The gas found in coal is also a source of energy which is being utilized to a great extent these days and this gas is known as coal bed methane, because methane is the dominant constituent in all the gases present in it. This is one of the most recent discoveries in past couple of decades as a promising energy resource.

Coal bed gas, which mainly consists of methane, has remained a major hazard affecting safety and productivity in underground coal mines for more than 100 years. Coal bed gas emissions have resulted in outbursts and explosions where ignited by open lights, smoking or improper use of black blasting powder and machinery operations. Investigations of coal gas outbursts and explosions during the past century were aimed at predicting and preventing this mine hazard. During this time, gas emissions were diluted with ventilation by airways *e.g.*, tunnels, vertical and horizontal drill holes, shafts and by drainage boreholes. CBM is a potentially important energy resource in many of the major coal mining countries of the world. Significant volumes of CBM are exploited worldwide with most of the gas originating from operational deep coal mines, and lesser quantities recovered from abandoned mine workings. Many coal-producing countries are now looking at the potential for wider application of CBM technologies to maximize the exploitation of gas from coal seams. CBM is a clean fuel with similar properties to natural gas when not diluted by air or other non-combustible mine gases. Thus, coal bed methane as a mining hazard was harnessed as a conventional gas resource.

Gas demand is rising sharply in India and CBM will compete with imported natural gas and liquefied natural gas (LNG) to meet these demands. India imported about 6.0 billion m^3 of natural gas in 2008. CIL estimates a supply gap of approximately 40 billion m^3 will eventually be reached that must be met by imports. CBM gas would compete favorably with imported coal, gas, or LNG on a fuel cost basis for power generation. There is thus an assured market for CBM, provided deliverability infrastructure is developed. Investment in coal and gas transportation infrastructure, including gas gathering, transportation and distribution, are necessary to fill this gap and to move CBM from coal fields to local and more distant end-use markets. End-use markets include rural power generation, commercial power generation, and transportation fuels. Limitations in cost and investment capital, however, remain significant barriers to technology development, application, and CBM project development in India.

CBM FORMATION AND CHARACTERISTICS

CBM is also known as natural gas from coal. It is created as a by-product of the coalification process during which organic materials are transformed into coal by pressure, temperature and time. Initially, plant and animal remains are turned into peat as they decay and undergo bacterial and chemical changes. That peat is gradually transformed into coal as it is buried by layers of sediment and subject to increasing pressure. The decomposition of the organic material during this process produces methane and other gases, which are adsorbed to the coal surfaces by Van der Waals forces created by natural pressure from the overlying rock and water within the coal formation. Coal is able to adsorb a significantly larger amount of gas than other types of source rock because of it large internal and external surface areas. CBM is classified as an unconventional natural gas because the coal bed acts as both the source of the gas and its storage reservoir, unlike conventional natural gas which is trapped within the pore space of rock distinct from where it was formed.

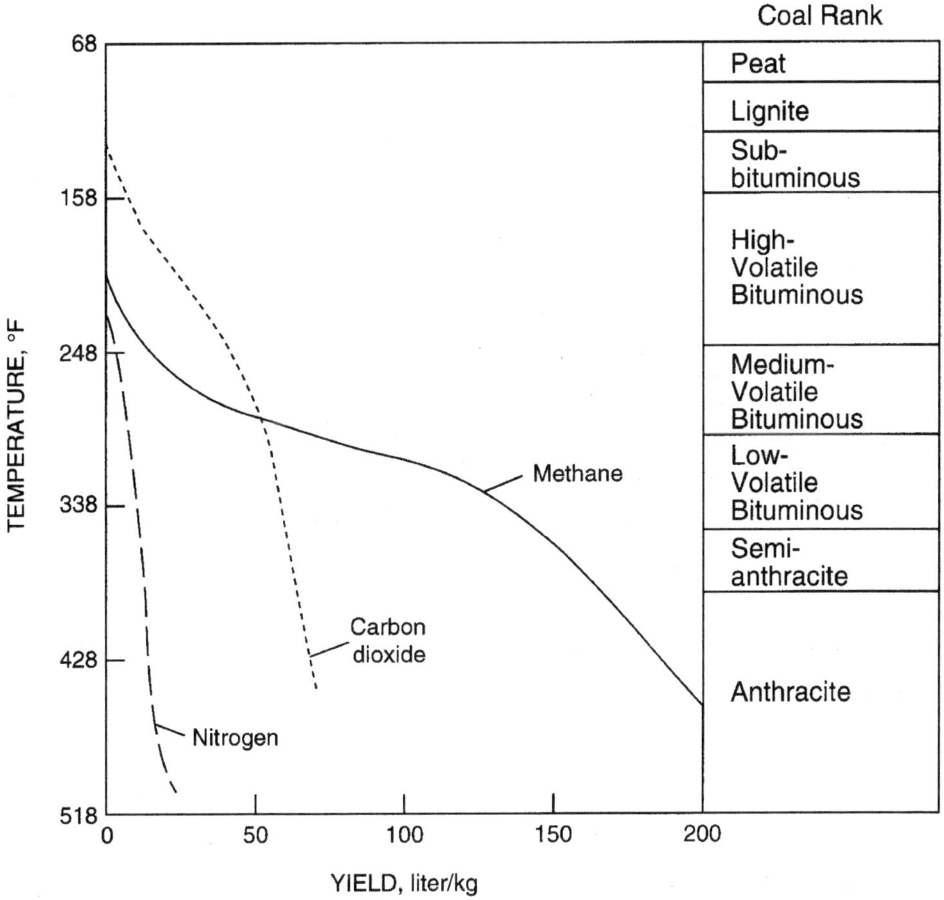

Figure 1: Gas Generation with Rank of Coal

CBM is found mainly adsorbed on the internal surface of coal. It is also found dissolved in associated ground water. CBM generation and in–place-gas content are deciding factors for CBM exploration. These are guided by various petrography and geological factors like adsorption capacity, depth, macerals content, maturity, cleat system, hydrogeologic condition, structural setup etc. In-place-gas and gas generation are also controlled by permeability, microspores, action of micro organism, reservoir pressure, hydrodynamics of the area, cleat aperture, width and extension, fault etc.

GEOLOGICAL CONTROLS ON CBM

Geology of the area affects and modifies mainly the generation, retention and transportation mechanism of coal seams. Therefore study of geological parameters on coal seams is very necessary for the development of CBM. There are a lot of factors that affects the CBM development. The important among these are as follows:

- *Depositional condition*: When higher concentration of plant materials is deposited in the basin, coals are formed. There may be two type of origin of coal; In- situ and drifted. In- situ coal contains biogenic gas as well as thermogenic gas while drifted coal contains thermogenic gas. With the help of isotope analysis one can know that whether the gas is thermogenic of biogenic.
- *Climate*: Warm and humid climate of sub- tropical region is most suitable for the formation of coal.
- *Structural activity*: If area is tectonically unstable then the coals are mostly faulted and jointed. Large scale faults are harmful because it provides path for the escape of gas from the coals.
- *Thickness and depth of coal*: Thick coal seams contain more gas content than the thin coal seams. As the depth of coal seams increases the pressure of overburden on the seams increases, due to which adsorption capacity of coal increases. Thus gas content are directly proportional to the depth of coal seams.
- *Rank of the coal*: Rank is the compositional maturity of the coal. High rank coals have higher maturity. The rank is based on volatile matter, fixed carbon, heating value, coking power, etc. High rank coals have higher adsorption capacity and vice versa.
- *Petrography of coal*: Coal is mainly composed of macerals. There are three maceral groups- Vitrinite, Liptinite and Inertinite. The vitrinite coals have more micropores containing higher gas quantity. Inertinite have least gas quantity.
- *Temperature and Pressure*: If temperature of the reservoir is high, then methane adsorption capacity is low and vice versa. On the other hand methane storage capacity increases with pressure.
- *Cleat system*: Cleats are the fractures in the coal seams. It provides path for the flow of gas within the coal seams. If the coals have higher porosity and permeability due to cleats, then gas can easily flow. Thus higher cleats density is profitable for the development of CBM.
- *Hydro-geological Condition*: Coals found at the lower and intermediate depths, are generally contains aquifer but coal seams found below seven hundred meters depth are mostly dry and absence of water. Good aquifers hamper the rate of production and also increase the duration of dewatering and finally the cost of production.

TYPE OF COAL BED METHANE

It is very important to distinguish different categories of methane from coal as described in Chapter 11 and explained below.

- *Coal Mine Methane (CMM)*: Methane is released as a result of mining activity when a coal seam is mined out and if not controlled to prevent the accumulation of flammable mixtures of methane in air (5-15 per cent) it presents a serious hazard. Gas drainage techniques are used to enable planned coal production rates to be achieved safely by reducing gas emissions into long wall mining districts to a flow that can be satisfactory diluted by the available fresh air. In well managed mines, in favourable geological and mining conditions, the methane concentrations in drained CMM can reach 70 per cent or more. CMM of such quality may be utilized. Methane capture and its utilization from coal mines is generally not practiced in India as current levels of coal production in gassy mines are generally achievable using ventilation controls but even where there may be some safety benefit there is some resistance to introducing gas drainage due to lack of technology, expertise and experience.
- *Ventilation Air Methane (VAM)*: Methane released from coal seams into the ventilation air of the active coal mine is called Ventilation Air Methane (VAM). Concentrations of methane in the ventilation air are generally, limited by law, for safety reasons, at 0.5 to 2 per cent in different parts of a mine with variations depending on the country. Concentrations can be controlled by the volume of ventilation air circulated (dilution) or through special drainage CMM. The concentration of methane in VAM is typically 0.8 per cent or less and is too low for conventional utilization purposes. However, technologies are being developed to remove the methane, and where additional gas is available to generate electricity using the thermal energy recovered.
- *Abandoned Mine Methane (AMM)*: When an active coal mine is closed and abandoned, methane continues to be emitted from all the coal seams disturbed by mining, decaying gradually over time unless arrested by flooding due to groundwater recovery. Depending on the methane concentrations, local regulations and the geology it may be possible, or required for public safety reasons to continue draining or venting AMM. No AMM schemes are in place in India and initial investigations show limited promise.
- *Virgin Coal Bed Methane (VCBM)*: Coal bed methane and virgin coal bed methane are terms conventionally used for methane drained and captured directly from coal seams. CBM is generally reserved to describe the gas produced from the surface bore holes ahead of mining for coal mine safety and coal production reasons. VCBM is produced by similar process but completely independent of mining activity. Methane concentration in VCBM is generally very high (around 90 per cent) and can be used as a replacement for natural gas supplies.

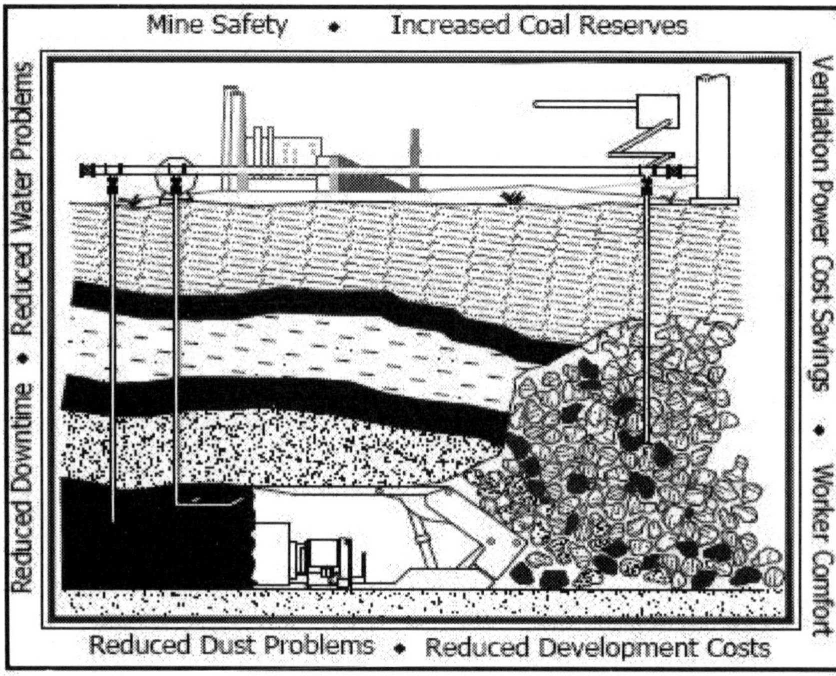

Figure 2: Coal Bed Methane Drainage

QUALITY AND POTENTIAL

Coal-bed methane is the world's cleanest burning fossil fuel. It is typically sweet gas with few impurities, and when produced is of such a high quality that very little processing (if any) is necessary before transportation through pipelines. Since methane gas is the primary component of CBM, it can be used interchangeably with conventional natural gas.

FACTORS INFLUENCING RECOVERABILITY

CBM recoverability from a particular coal bed is determined by the permeability of the seam and the volume of gas it contains. Permeability is affected by the natural network of interconnected fractures present in the coal and created by subsurface compression forces. These fractures are referred to as "cleats" and allow for the movement of water and gas to the well bore during production. Cleat widths are generally greater near the surface but decrease with depth as pressure is increased, reducing permeability. The type of coal present in a particular deposit depends on the conditions that created it. Coal is classified by rank, which is directly related the amount of pressure it was subject to during its creation and how long it was exposed to that pressure. Higher-ranked coal contains more carbon and energy-producing potential as well as greater volumes of CBM. Higher rank, however, indicates higher formative pressures which adversely affects permeability. This inverse relationship between permeability and CBM content makes production difficult, as producers must identify coal seams that are permeable enough to permit production but contain sufficient CBM to be economical.

PRODUCTION TECHNIQUES

The adsorption capacity of coal increases with temperature and pressure, but decreases where other elements such as water or carbon dioxide are present and displace gas from adsorption sites that it would otherwise occupy. Production therefore requires altering the pressure within the coal seam to cause the CBM to desorb and migrate to a well bore, which permits the flow of gas to the surface where it can be compressed for transportation through pipelines. This can be accomplished by removing water from the coal bed which reduces ambient pressure and allows the CBM to desorb, or by injecting carbon dioxide into the coal which displaces the methane from adsorption sites. If natural fissures are not present or do not allow for sufficient flow of gas during production, the coal bed must be fractured by pumping fluid down the well at high pressure to create a network of cleats that allow for gas movement (US EPA, 2004). A popping agent such as sand is added to the fracturing fluid to prevent these fractures from closing when the fluid pumping stops.

METHODOLOGY OF RESOURCE ESTIMATION

Stored gas volume or gas-in-place generally refers to the volume of gas that is currently stored within coal seams. The current stored volume of gas is dependent upon two primary conditions: a) the volume of gas within the coal matrix and fracture system, and b) the volume of the coal seam. Two methods have been suggested for estimation of the volume of gas that is currently contained within a coal seam: volumetric and material balance. The volumetric method is, by far, the dominant method in use today for the estimation of coal bed methane resources, primarily because of the ease of use and the less rigorous data requirements. The use of the material balance method, on the other hand, requires a significantly greater understanding of the coal seam reservoir, which results in the infrequent use of this method for coal seam gas resource and reserve estimation. The estimation of the volume of gas that is contained within a coal seam or seams and the estimation of the volume of gas that could be recovered from that seam is dependent upon numerous physical variables.

CBM DEVELOPMENT IN INDIA

Compared with the major coal mining countries, India has relatively modest CBM resources. Nevertheless, the government of India considers VCBM and CMM as a potentially important clean energy source. As within many countries ownership, exploration and extraction of CBM in India does not fall under one regulatory body. While the responsibility for future energy needs, including CBM, falls under the Ministry of Petroleum and Natural Gas, CBM where coal mining activities are taking place the responsibility falls under Coal India Ltd (CIL) and the Ministry of Coal.

- *VCBM*: Geological appraisal has identified about 20,000km^2 of coalfield areas with a VCBM potential in which recoverable gas reserves are estimated at 0.8 TCM. The bituminous coal basins with VCBM potential are: Damodar-

Rajmahal in West Bengal and Bihar, Sone-Mahanadi, and Narmanda-Pranhita-Godavari in Madhya Pradesh, Orissa, Andhra Pradesh, and Maharashtra. Tertiary lignite-bituminous coal basins with CBM potential include Cambay in Gujarat, Barmer in Rajasthan and Cauvery in Tamil Nadu. The Directorate General of Hydrocarbons, Ministry of Petroleum and Natural Gas is responsible for establishing the policy framework for VCBM development in India and has evolved a model contract to facilitate global bidding. Exploration will be licensed under a concession agreement, different from the production sharing contract approach used in China. The government offers tax breaks, freedom to sell the gas and provisions for 100 per cent cost recovery. A royalty will be paid on produced gas.

- *CMM*: Most underground coal mines (90 per cent) use room-and-pillar methods of extraction. Gas drainage will therefore mainly involve pre-drainage of the worked seam although there may be a possibility of establishing post drainage where pillar recovery is practiced. CIL through the Central Mine Planning and Design Institute (CMPDI) and Central Institute of Mining and Fuel Research in association with the UNDP and GEF are involved in a demonstration project on CMM recovery and utilization. The aim of the project is to demonstrate the commercial feasibility of using CMM extracted during coal mining for power generation and as an alternative fuel for vehicles. Lack of technical know-how in India is considered a barrier to effective use of CMM. It is anticipated that the project will encourage adoption of drilling technology and working practices to drain and use methane more effectively. Two mines, Moonidih and Sudamdih in the Jharia coalfield are involved.
- *AMM*: The potential for AMM has yet to be assessed. India has a shortfall in energy supply but CBM is only likely to be able to make a modest contribution at best.

CMM AND CBM STATUS IN THE WORLD

Underground coal mines worldwide liberate an estimated $29 - 41 \times 10^9$ m^3 of methane annually, of which less than 2.3×10^9 m^3 are used as fuel. The remaining methane is emitted to the atmosphere, representing the loss of a valuable energy resource. Coal mine methane recovery and use represents a cost-effective means of significantly reducing methane emissions from coal mining, while increasing mine safety and improving mine economics. The world's ten largest coal producers are responsible for 90 per cent of global methane emissions associated with the coal fuel cycle. China is the largest emitter of coal mine methane, followed by the Commonwealth of Independent States, or CIS (particularly Russia, Ukraine and Kazakhstan), the United States, Poland, Germany, South Africa, the United Kingdom, Australia, India and the Czech Republic. Most of these countries use a portion of the methane that is liberated from their coal mines, but the utilization rate tends to be low and some countries use none at all. Methane is used for heating and cooking at many mine facilities and nearby residences. It is also used to fuel boilers, to generate electricity, directly heat air for mine ventilation systems and for coal drying. Several mines in the United States sell high-quality mine gas to natural gas distributors. There are barriers to decreasing methane emissions by increasing coal mine methane use. Many of the same barriers are common to a number of the subject countries. Technical barriers include low-permeability coals, variable or low gas quality, variations in gas supply and demand and lack of infrastructure. Economic and institutional barriers include lack of information pertinent to development of the resource, lack of capital and low natural gas prices. A possible option for encouraging coal mine methane recovery and use would be international adoption of a tradable permit system for methane emissions.

The major CBM basins in the U.S. include: the San Juan, Piceance/Uinta, Raton, greater Green River and Powder River Basin. Between 1989 and 1999, total U.S. CBM production increased from 91 billion cubic feet to 1.3 trillion cubic feet (Tcf). In 1997, U.S. Geological Survey (USGS) estimates total U.S. coal bed methane resources at 700 trillion cubic feet (Tcf) with less than 100 Tcf economically recoverable (USGS Fact Sheet FS-019-97). To put these numbers in perspective, U.S. annual consumption of natural gas is 23 Tcf and rising. Industry and CBM opponents fiercely contest the technically and economically recoverable estimates. Conversely, in the San Juan basin where extensive CBM development has been completed, industry has quantifiable and site-specific data backing their 43 to 49 Tcf methane estimate and development costs. More recent estimates have placed San Juan basin reserves at 84 Tcf, with 8.5 Tcf of the 12 Tcf of recoverable gas already extracted. Methane potential in Colorado's Piceance basin, subject to less exploration and development than the San Juan basin, exhibits a wider range from 36 to 84 Tcf.

ECBM RECOVERY AND CO_2 STORAGE

The recovery of coal bed methane can be enhanced by injecting CO_2 in the coal seam at supercritical conditions. Through an in-situ adsorption/desorption process, displaced methane is produced and the adsorbed CO_2 is permanently stored. This is called enhanced coal bed methane (ECBM) recovery and it is a technique under investigation as a possible approach to the geological storage of CO_2 in a carbon capture and storage system.

Carbon dioxide sequestration in coal and enhanced coal bed methane recovery have recently become a vibrant research areas due to their potential for permanently storing carbon dioxide in deep and unminable coal seams while producing methane to offset the costs. The field applications of this technology have been demonstrated in North America, Asia and Europe in medium- to large-scale tests. Future smaller scale demonstration tests are also being planned in different countries to further explore the potential of the technology to reduce atmospheric CO_2 levels and to investigate various operational problems and unexpected outcomes that have been encountered in the previous field tests. Such problems encountered during coal seam CO_2 sequestration are largely due to the heterogeneous nature of coal and the interactions of gases with different structural and chemical features of the coal. It has been concluded that these interactions change the reservoir behavior of coal seams as well as fluid flow and storage characteristics. The general consensus among geologists, engineers, coal scientists and energy economists working on different aspects of CO_2

Figure 3: Methane Recovery by CO_2 Injection

sequestration is that there is a lack of adequate understanding of how coal–gas interactions can change reservoir and flow behavior, and how better analytical and numerically methods can be developed to characterize these changes.

ENVIRONMENTAL CHALLENGES AND CBM RECOVERY

The production of coal bed methane is accompanied by significant environmental challenges, including prevention of unintended loss of methane to the atmosphere during underground mining, and disposal of large quantities of water, sometimes saline that are unavoidably produced with the gas, especially in the early stages of production. While economic quantities of methane can be produced water disposal options that are environmentally acceptable and yet economically feasible are a concern. Water may be discharged on the surface if it is relatively fresh, but often it is injected into rock at a depth where the quality of the injected water is less than that of the host rock.

FURTHER READING

[1] Bryner, Gary, 2002, Coal bed Methane Development in the Intermountain West: A Primer, Natural Resources Law Center, University of Colorado School of Law. Bureau of Land Management (BLM).

[2] Choate, R., C.A. Johnson and J.P. McCord, 1984, Geologic overview, coal deposits, and potential for methane recovery from coal beds-Powder River basin. In: C.T. Rightmire, G.E. Eddy and J.N. Kirr (Eds.), Coal Bed Methane Resources of the United States, Am. Assoc. of Petrol, Geol., Studies in Geology Series 17, pp. 335–351.

[3] Coal Evolution, ibid. at 2; The Coal Resource, supra note 1 at 4.

[4] Jeffrey R. Levine, 1990, University of Alabama, Generation, Storage and Migration of Natural Gas in Coal bed Reservoirs Lecture Notes for Short Course, Alberta Research Council (August 20-24, 1990) at 111.

[5] King, G.R., 1993, Material-Balance Techniques for Coal-Seam and Devonian Shale Gas Reservoirs with Limited Water Influx. SPE Reservoir Eng., February 1993, pp. 67–72.

[6] Nowak, Henry C., 1991, Depositional Environments and Stratigraphy of Mesaverde Formation, Southeastern Piceance Basin, Colorado-Implications for Coal bed Methane Exploration, in S.D. Schwochow, D.K. Murray and M.F. Fahy, eds., Coal bed Methane of Western North Americam, Denver, CO: Rocky Mountain Association of Geologists.

[7] USEPA, Evaluation of Impact to Underground Source of Drinking Water by Hydraulic Fracturing of Coal Bed Methane Reservoirs, 2004, United States Environmental Protection Agency, online: U.S. Environmental Protection Agency <www.epa.gov> at 1-3 [U.S. EPA].

[8] Watson, R.W., 2002, The Growing Natural Gas Supply and Demand Imbalance: the Role that Public Lands and Federal Submerged Lands Could Play in the Solution. Assistant Secretary for Land and Minerals Management, U.S. Department of the Interior– Congressional Testimony before the Resources Subcommittee on Energy and Mineral Resources, U.S. House of Representatives, July 16, 2002.

[9] Zuber, M.A., 1996, Basic reservoir engineering for coal, in Saulsberry, J.L., Schafer, P.S., Schraufnagel, R.A. (Eds.), A Guide to Coal Bed Methane Reservoir Engineering. Gas Res. Inst., Chicago, IL, pp. 3.1–3.34.

Chapter 13

Oceans: A Ready Solution or a Last Frontier in Carbon Dioxide Mitigation?

Nittala S. Sarma

Marine Chemistry Laboratory, School of Chemistry, Andhra University, Visakhapatnam – 530 003

SUMMARY

Oceans cover ~71 per cent of the planet Earth's surface area. The lowest estimate of CO_2 that can be stored in the Oceans is ~5,000Gt, *i.e.*, CO_2 released at the present rate in the next 1,000 years can be dispensed with. Adding cheap and harmless chemicals that make the seawater further basic, *e.g.*, lime stone ($CaCO_3$) and lime ($Ca(OH)_2$) would cause increased absorption of the acidic CO_2 gas by seawater. This may, however, have adverse effect on the marine ecosystem and fishery resources. A more intensely pursued option is Ocean Iron Fertilization (OIF). Credited with a biological 'must' requirement and characteristic redox chemistry, Fe is in deficit in the nutrient rich but productivity poor waters that surround the Antarctica extending over an area equivalent to ~20 per cent of the global oceanic surface. If the cheap Fe were to be supplied by human intervention, these waters would fix as many as ~3000 moles of CO_2 for every Fe atom. Brief history of experiments of ocean iron fertilization conducted so far, the results of these and LOHAFEX conducted in Southern Ocean, besides the legal and ethical issues of plankton ecology are described.

INTRODUCTION

At present, various options are being considered for the mitigation of carbon dioxide in the atmosphere:

1. Shifting from carbon based fossil fuels to alternative energy sources: wind power, nuclear power, solar power, renewable energy, electric or hybrid automobiles, fuel cells, etc.,
2. Carbon capture and storage in deep geological layer formations and on the ocean bed,
3. Carbon sequestration by plants on land and in the ocean
4. Enhancing natural CO_2 sinks, *e.g.*, doubling the photosynthesis efficiency
5. Cooling the Planet by creating a solar sunshade to deflect ~10 per cent of solar radiation.

Before the oceans are considered as a viable option for CO_2 sequestration, an overview of the role of oceans in CO_2 equilibria needs to be considered.

ROLE OF OCEANS IN CARBONATE EQUILIBRIUM

The oceans contain 97.8 per cent of the global water. And this water contains ~2.5 mmoles of dissolved $CO_2\ l^{-1}$, present in the form of bicarbonate (90 per cent) and carbonate (10 per cent) ions. The ionic balance is ensured by their association with stronger cations such as calcium and magnesium, a reason why natural seawater is basic (pH~8). This equilibrium ensures that the seawater also acts as a buffer, *i.e.*, small changes in the flux of CO_2 shall cause little variations in pH. It sounds good, but the effect is seen always. For example, in the deep horizons where organic matter undergoes oxidation to CO_2, the pH drops down from 8.2 in surface water to as low as 7.8. A similar thing happens if surface seawater is in contact with an atmosphere of increased CO_2 concentrations. The fall of pH and the associated coral bleaching are well documented phenomena.

There is an intense coupling between the atmospheric CO_2 and oceanic CO_2. Of the 7.4 Gt/y of CO_2 liberated into the atmosphere, oceans absorb as much as 2 Gt, and leave 3.4 to accumulate in the atmosphere. An increase by 1 ppmv means an accumulation in the atmosphere of CO_2 of ~2 Gt, *i.e.*, since 1950, the oceans have thus gained 120 Gt. The fate of the remainder 2 Gt of CO_2 is largely unknown.

Algal Cultivation

Algal cultivation is supposed to play an important role in carbon sequestration. At the Tokyo University of Marine Science and Technology, scientists are reported [1] to try setting up huge water-borne farms with 100 vast nets each measuring 100 km^2 (10x10 km) each floating off the North East coast of Japan. Each net will grow seaweed biomass of 270 kilotons yr^{-1} by species rotation and absorb 36 tons of CO_2. This will also provide food for marine life. The nets can be tracked by GPS.

Ocean Injection and Storage

The ocean already contains ~140,000 billion tonnes of CO_2 compared to annual worldwide anthropogenic emissions of 27 billion tonnes (2004 figure: Source: Energy Information Administration, The Guardian).

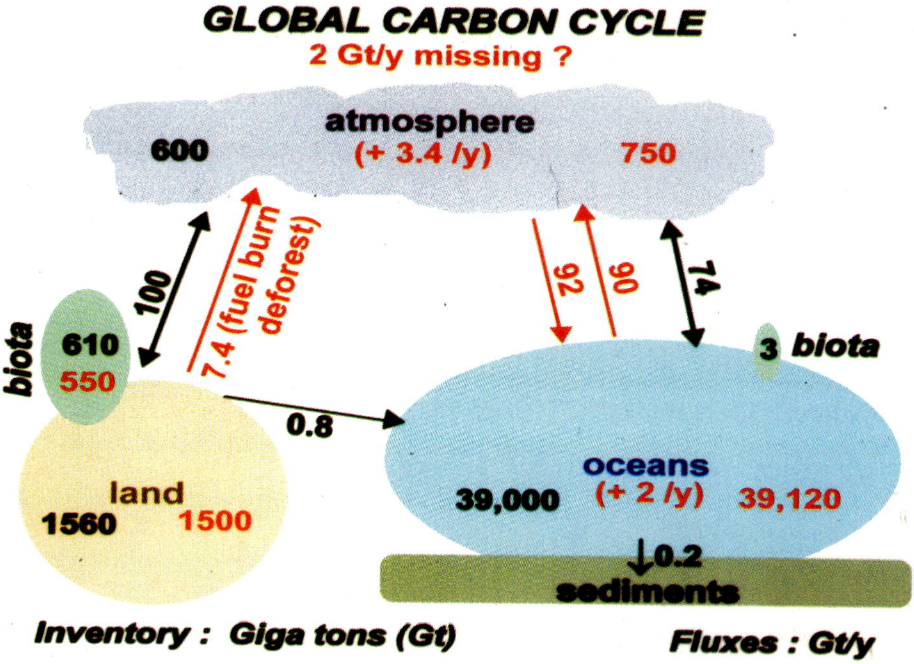

Figure 1: Global Carbon Cycle (Inventory: Gt; Fluxes: Gt yr^{-1}):
(Nos. in black are values for 1950; Nos. in red are present day values,
12 g C≡ 44 g CO_2)

Dissolution Type

Deep injection of CO_2 at depths greater than 1000 meters would make it dissolve. Marchitti [2] first conceived that piping CO_2 into the outflow of the Mediterranean Sea would have the CO_2 transported along with the Mediterranean water to deeper depths of the Atlantic. Increasing CO_2 in seawater causes acidification that can cause noticeable ecosystem changes such as bleaching of corals. It will also reduce the ability of seawater to take up more CO_2 as carbonate ion of the carbonate system is transformed to bicarbonate ion:

$$CO_3^{-2} + CO_2 + H_2O \rightarrow 2HCO_3^-$$

Recently, NOAA, in collaboration with a few other institutions has launched the first-of-kind buoy to monitor acidification in the Gulf of Alaska (North Pacific) [3].

Lake Type

CO_2 occurs as liquid under supercritical conditions, *i.e.*, when the temperature is below 23ºC and pressure high. At great depths of the ocean, these conditions are reached easily. At a depth of 3000 meters, the T is ~4ºC and P ~300 bar, and any CO_2 injection, on equilibration with ambient conditions would liquify, and since liquid CO_2 is denser than seawater, would settle on the seafloor in the form of a lake bounded between topographic highs [4].

Converting CO_2 to Bicarbonates

Adding limestone to a solution of CO_2 produces bicarbonate.

$$CO_2 + CaCO_3 + H_2O \rightarrow Ca(HCO_3)_2$$

Little is known about the potential environmental consequences to ocean ecosystem and natural biogeochemical cycles.

OCEAN IRON FERTILIZATION (OIF)

For primary production to take place in the illuminated surface waters of the sea, usually, nutrient elements, *e.g.*, nitrogen, phosphorus and silicon, in the forms of nitrate, orthophosphate and orthosilicate are the limiting factors as all other nutrient elements are present in large excess over the needed concentrations. Although this is true of major constituents like CO_2, Ca, K, Mg, Sr, Na etc. and even trace components like Cu, Zn, Co, Ni, V, Mo etc., Fe, because of its characteristic chemistry behaves exceptionally.

Uniqueness of Iron

Iron is an important nutrient metal in biological systems due to its 'd' orbital electronic configuration facilitated by *redox* transitions that take place at physiological redox potentials. Fe is suited to many enzyme and electron carrier proteins. Nitrate reductase involved in the assimilation of CO_2 (reduction of nitrate to nitrite) and nitrogenase enzyme involved in N_2 fixation by BG algae in the ocean surface water

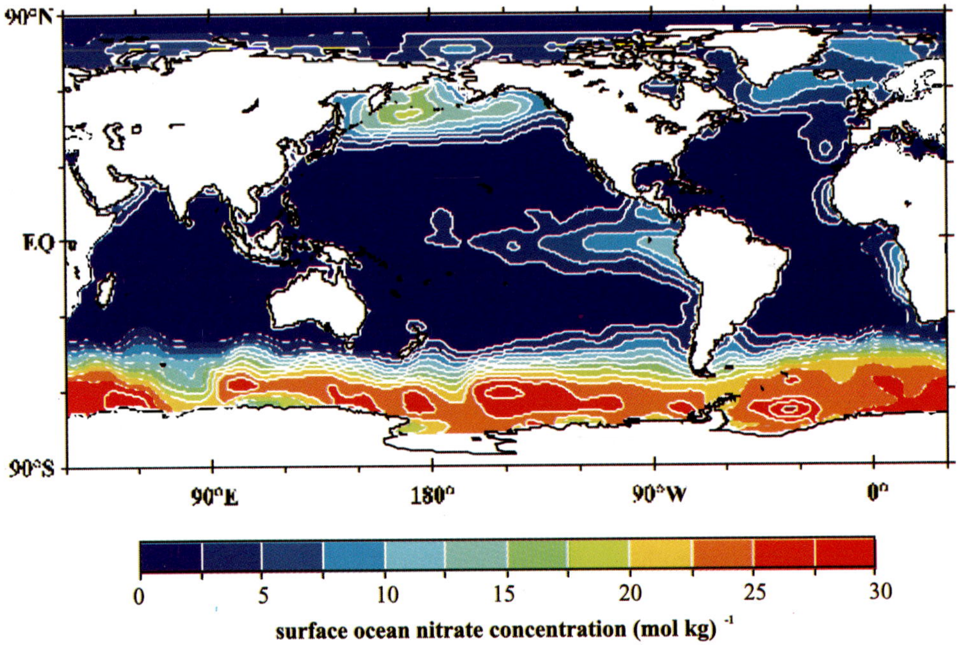

Figure 2: Surface Ocean NO_3^{-1} Concentration (μM kg^{-1}). HNLC waters have NO_3^{-1}>20 μM kg^{-1}

are Fe requiring enzymes. The known sources of Fe in the ocean are (i) air-borne terrigenous dust and magmatic Fe from volcanic eruptions, and (ii) upwelling of the terrigenous and magmatic sourced Fe. The input of Fe into the ocean has changed in the geological past and had profound impact on climate change. Due to its redox and solution chemistry mentioned and particle reactivity, its dissolved concentration is low (<20 picomolar). Unless there is an ongoing supply, Fe can become a limiting nutrient, even though under normal conditions, it is the other nutrients N (and P) that are the ones that control primary production due to their large requirement for phytoplankton. Hence, adding Fe to HNLC waters promotes phytoplankton blooms and a CO_2 draw down from the atmosphere due to a "gas equilibrium" at the water-air interface. Indeed, high levels of Fe caused by intensified storms are linked to Ice Ages in the geological record. This led Martin, who introduced this concept [5] to fancy: "give me half a tanker of Fe and I will give you next Ice Age"[6]. Fe fertilization may cost only 50-100 times lower than other proposed sinks. Ritschard calculated [7] that carbon sequestering potential of Fe fertilization and creation of large macroalgal (kelp) farms is 0.7-3.0 $GtCyr^{-1}$ from the atmosphere at a cost of US$ 5-300 $t^{-1}Cyr^{-1}$.

HNLC (high nutrient low chlorophyll) waters typically have NO_3>2μM, and Chl a <0.5μg. They occur in the Subarctic northeastern Pacific, the eastern equatorial Pacific and the Southern Ocean water covers the ocean surface to an estimated 20 per cent extent HNLC waters (Figure 2).

Every mole of Fe, can promote fixation, on average of about 3,000 moles of CO_2. This ratio could in fact be as high as 10,000 and depends on ecological factors, chief amongst them being the type of organisms produced: diatoms, flagellates etc. and species. The size of the iron particles is critical; particles of 0.5~1μ or less seem to be ideal both in terms of bioavailability and sink rate. The dosage required to fertilize a bloom is 0.05 g m^2 for a 50 m deep mixed layer. For keeping Fe in solution form, the preferred form is a Fe chelate with lignin acid sulfonates. All is not well until the fixed organic carbon is transported to the deep, as otherwise cycling within the water column would liberate back CO_2. The time scales are of course variable: if metabolism takes place within the upper 100m, the time scale is a few years; below the mixed layer it would gradually increase to as much as a few thousand years at the bottom of the sea. The fraction that survives recycling and reaches the seafloor, in the world oceans, is only 1-2 per cent of the surface fixed carbon. Of this, only half is preserved in the sediment. Compared to this, nearly 20 per cent of the surface fixed carbon, promoted by Fe fertilization has been found to reach the seafloor, although the values are variable between different experiments.

The proof-of-concept meso scale Fe fertilization experiment was first conducted in 1993 off Galapagos islands, and since then as many as ten large scale experiments have been conducted in different sectors of the Southern Ocean including the last one in the Atlantic sector by Indian and German scientists during January-March, 2009.

- ☆ Ironex II, 1995
- ☆ SOIREE (Southern Ocean Iron Release Experiment), 1999
- ☆ EisenEx (Iron Experiment), 2000

- ☆ SEEDS (Subarctic Pacific Iron Experiment for Ecosystem Dynamics Study), 2001
- ☆ SOFeX (Southern Ocean Iron Experiments– North and South), 2002
- ☆ SERIES (Subarctic Ecosystem Response to Iron Enrichment Study), 2002
- ☆ SEEDS-II, 2004
- ☆ EIFEX (European Iron Fertilization Experiment), 2004
- ☆ CROZEX (CROZet natural iron bloom and Export experiment), 2005
- ☆ LOHAFEX, 2009

CHALLENGES TO *OIF*

There are some key issues that cause hurdles on the way to implementing OIF on a large scale:

(*i*) The Southern Ocean (SO) is the only ocean region where substantial amounts of the phytoplankton nutrients, nitrate and phosphate are supplied to the surface waters (by upwelling of deep ocean water in the South and returned to the deep ocean by downwelling along its northern fringe).

(*ii*) In Fe fertilization, an artificial chelator lignin acid sulfonate is added to keep Fe dissolved and bioavailable. In nature, the membrane bound siderphores do the trick. This causes pollution with the polymer that is not easily biodegraded.

(*iii*) Community shifts, and appearance of HAB (harmful algal blooms).

(*iv*) Deep ocean hypoxia resulting in the formation of CH_4, whose global warming potential is 24 times more, and N_2O that is 206 times as powerful.

(*v*) Original estimates for the Southern Ocean suggest that as much as 1.8 billion tons of C could be removed annually. But, recent models predict that if all the nutrients of Southern Ocean waters were converted to organic carbon over the next 100 years (extremely unlikely!), only 15 per cent of the anthropogenic CO_2 can be sequestered[8]. Hence, this is not a permanent solution.

(*vi*) By ocean fertilization, the earth's carbon cycles are redrawn on time scales never happened in the geological past with potential risks apprehended to outweigh beneficial effects.

(*vii*) Recently, it is shown that it is light rather than Fe that controls photosynthetic production in Southern Ocean phytoplankton populations during austral autumn[9].

(*viii*) Increased production does not ensure increased export of carbon, as much of the biomass undergoes metabolic degradation back to CO_2 within the euphotoc zone. Deep POC fluxes (>2000 m) are higher with carbonate-dominated particles, even of lower surface concentration than with diatom (silica) dominated particles [10]. However, when ballasting minerals are abundant in this region, POC export to depths is favoured.

(*ix*) In deed, in the Southern Ocean, HBLE (high biomass low export) regimes are common as observed recently [11] when different size fractions of particulate organic carbon of the twilight zone (100-1000 m) that over which the euphotic layer sits are analyzed.

(*x*) le Quere *et al.* [12] recently argue that the Recent climate change has caused saturation of the Southern Ocean CO_2 sink due to wind that became more turbulent then in response to global warming.

Further, as of now, carbon trading applies to terrestrial sinks, *e.g.*, reforestation only. More over, it is illegal to seed the ocean waters with ship-loads of Fe as it can be termed 'dumping' under the existing laws. Even so, several patents have been filed on ocean fertilization anticipating a global market.

Because of these issues, it is suggested[8] that ocean fertilization in the open seas and terrestrial waters should never become eligible for carbon credits[13].

But the oceans seem to hold much hope for their deep basins. The IEA Greenhouse Gas R&D programme estimates that the world-wide sequestration potential of the geological formations to be highest for the deep ocean [14], compared to deep saline aquifers and other geological formations (Table 1).

Table 1: CO_2 Sequestration Potential in Different Reservoirs

Reservoir	*Potential (billion tons of CO_2)*
Deep Ocean	5,100- >100,000
Deep Aquifers	320–10,000
Depleted gas reservoirs	500-1100
Depleted oil reservoirs	150-700

In a high CO_2 environment, increasing concentration of CO_2 may have an unexpected welcome sympathy in enhanced CO_2 uptake. In *Trichodesmium*, a diazotrophic, *i.e.*, N_2 fixing Cyanobacteria over large parts of the ocean, CO_2 concentrations have been observed to affect cell growth; the POC increasing without altering the C/N ratio, but increasing the C/P (and N/P) ratio. This process has the potential of converting the nitrate (and Fe) limitation to one of phosphate limitation, a welcome feature!

The $CaCO_3$/POC production ratio is one of the key biotic climatic variables. In hexa corals (hard corals), the ratio reaches highest: 10^2-10^4. Among the primary producers, it is high in coccolith algae that are also the largest contributor (as much as ~30 per cent) to global ocean primary production. Its calcification ratio ($CaCO_3$/POC) is light-dependent. It first increases in the euphotic column until the light irradiance becomes a limiting factor, but there after, it is decreased. Studies show that, by developing huge coccolith ponds that operate under optimized conditions, CO_2 can be sequestered into lime deposits.

CONCLUSION

CONCLUSION

An effective formulation of a dynamic green ocean model that incorporates the key plant functional types (PFT): diatoms, nano and picoplankton, N_2 fixers, coccolithophorids, and phaeocystis (dimethyl sulfide producer) and their ecology is necessary before any large scale ocean fertilization is implemented, and the international scientific community is indeed alive to this requirement.

REFERENCES

[1] http://www.yomiuro.co.jp

[2] Marchetti, C., 1977, Climate Change, 1, 59.

[3] NOAA Press release, 12 June 2007.

[4] Nobel Intent: September 19, 2006, Carbon dioxide lakes in the deep ocean, posted by John Timmer

[5] Martin J.H. *et al.*, 1988, Nature, 331, 341.

[6] Chisholm S.W. and F.M.M. Morel, 1991, Limnol. Oceanogr., 36, 1507.

[7] Ritschard, R.L., 1992, Water, Air and Soil Pollution, 64, 289.

[8] Chisholm, S.W., *et al.*, 2001, Science, 294, 309.

[9] van Oijen, T. *et al.*, 2004, J. Plankton Res., 26, 885.

[10] Francois *et al.*, 2002, Global Biogeochemical Cycles, 16, 1087.

[11] Lam, P.J. and K.B. Bishop, 2007, Deep Sea Res.II, 54, 601.

[12] LeQuere, C.L. *et al.*, 2007, Science, 316, 1735.

[13] Ormerod W., 1994. The disposal of carbon dioxide from fossil fuel fired power stations, IEAGHG/SR3, IEA Greenhouse Gas R&D Programme, Cheltenham, UK.

[14] Ramos, B. *et al.*, 2007, Global Biogeochemical Cycles, 21, GB2028, doi: 10.1029/2006GB002898.

Chapter 14

Soil and Vegetation Carbon Pool and Sequestration in the Forest Ecosystems of Manipur, NE India

P.S. Yadava

Department of Life Sciences, Manipur University, Imphal – 795 003

SUMMARY

Human activities have affected the world climate through the increase in the greenhouse gases concentration in the atmosphere mainly CO_2. To investigate the effect of atmospheric CO_2 it is essential to evaluate carbon sequestration and storage in the different ecosystems. Different soils store large number of organic carbon and are globally important source and sink of the greenhouse CO_2. The paper deals with the assessment of carbon pool and sequestration in the soil and vegetation in the forest ecosystem of Manipur, N.E. India. The soil carbon stock ranged from 27.73 to 48.03 t C ha^{-1} and emission of carbon dioxide from the soils varied from 1.697 to 4.462 t CO_2 ha^{-1} yr^{-1} in different forest ecosystems. The carbon stock in the vegetation ranged from 9.89 t C ha^{-1} to 295.50 t C ha^{-1} in the different forest types of Manipur. Soil carbon pool and emission of CO_2 are highly influenced by vegetation types and environmental factors.

INTRODUCTION

According to recent IPCC report, there is strong evidence that human activities have affected the world climate through the increase in the greenhouse gases concentration in the atmosphere. Forest ecosystem acts as a CO_2 sink when there is an increase in the carbon stock retained in forest vegetation itself as well as in the form of organic carbon in the soil. However, deforestation and burning of forest release CO_2 in the atmosphere. Tropical forests have the greatest potential to sequester carbon primarily through reforestation, agro forestry and conservation of existing forest [1].

In North Eastern India, forest ecosystems have been a source and sink of carbon owing to shifting cultivation and deforestation. However, in the recent years the ban of extraction of woods in the forest by the Government of India, for commercial purpose, there may be increase in the forest cover and subsequent increase in the sequestration of carbon.

There are limited information on the carbon stock and rate of carbon sequestration on the Indian forests [2,3]. Therefore an attempt has been made to examine the soil and vegetation carbon stock and rate of carbon sequestration in the forest ecosystems of Manipur, North East India.

Manipur is a small isolated hilly state located on the eastern arm of the Himalaya the Purvanchal [4], which separate India from Myanmar in North East India along the Indo-Myanmar border lying north of the Tropic of Cancer. It is located between 23°80' N to 25°68' N Latitude and 93°05' E to 94°78' E Longitude. The state is almost square in shape with a fertile alluvial soil, central Imphal valley comprises only 10 per cent of total geographical area of 22, 327 sq. km and 90 per cent hilly area with population of 23,88, 634 [5].

The climate of the state is monsoon type with warm moist summer and cool dry winter. The mean maximum temperature varied from 24.16°C (January) to 35.9°C (May) and mean minimum temperature ranged from 4.5°C (January) to 23.1°C (August). The mean monthly rainfall ranged from 4.79 mm (January) to 195.88 mm (July). The mean annual rainfall is 1244.99 mm. The relative humidity ranged from 61.5 per cent (February) to 82.8 per cent (July). The year is divisible into three seasons *i.e.*, moist summer (March to May) and rainy (June to October) and winter (November to February) seasons.

FOREST VEGETATION

The natural vegetation of the State mainly consists of forests which occupies about four fifth of total geographic area, which is broadly classified as wet temperate forest, pine forest, wet hill forest, semi evergreen forest, teak *Gurjan* forest, bamboo brakes and grass brakes (Table 1).

The research on carbon stock and rate of sequestration of forests of Manipur is relatively few which was carried out at the Ecology laboratory, Department of Life Sciences, Manipur University, Imphal. The estimate of carbon storage and sequestration in the soil and vegetation are based on our studies carried out in recent years in different forest types of Manipur, North-East India [6-10].

Table 1: Areas Under Forest Types of Manipur

Sl.No.	Forest Types	Area (sq. km)	Percentage of Total (per cent)
1.	Wet temperate forest	1417.12	8.23
2.	Pine forest	2386.55	13.86
3.	Wet hill forest	8850.56	51.40
4.	Semi-evergreen forest	631.93	3.67
5.	Teak-Gurjan forest	599.22	3.48
6.	Bamboo brakes	3194.12	18.55
7.	Grass brakes	141.19	0.81
	Total	17220.69	100.00

Source: Forest Survey of India, 2003.

CARBON STOCK OF FOREST ECOSYSTEM

The total land area under forest is 17,220.69 sq. km. under different forest types. The forests comprise of natural primary forest in certain pockets and mainly fall under open forest and partially degraded to bamboo and grassland. The forest in North Eastern India are under biotic pressure mainly owing to deforestation for timber, fuelwood and shifting cultivation. The teak-Gurjan forest comprises a small area of 599.22 sq. km but harbors rich biodiversity.

However limited data are available on the carbon stock of the forest ecosystem on oak, pine and *Dipterocarpus* forests of Manipur (Table 2). The aboveground biomass ranged from 130.22 t ha^{-1} to 255.04 t ha^{-1} in oak forests, 173.47 (16 yrs old) to 591.00 t C ha^{-1} in Pinus plantation forest and 18.78 to 21.92 t ha^{-1} in secondary *Dipterocarpus* forest. The carbon stock in the present study is based on 50 per cent carbon content of above ground biomass [11]. The carbon stock in Oak forest ranged from 65.11 to 127.52 t C ha^{-1}, 86.73 (16 yrs old) to 295.50 t C ha^{-1} (40 yrs old) in Pinus plantation forest and 9.89 to 10.96 t ha^{-1} in secondary *Dipterocarpus* forest.

It shows that carbon stock increases with increase of stand age in pine forest plantation. Thus 40 years old Pinus plantation forests have higher carbon stock than *Dipterocarpus* forests. The carbon stock in the present *Dipterocarpus* forest is lower than *Dipterocarpus* forest of Philippines but carbon stock in Pinus plantation is higher than Pinus forests of Philippines [12]. The lower value in the present *Dipterocarpus* forest is due to a secondary young forest and though dominated by fast growing species of *Dipterocarpus tuberculatus* but subjected to logging and shifting cultivation[13].

At present Government of Manipur initiated plantation of tree in the degraded forests as well as under private lands. There is a ban on logging in the protected forest in the country which may lead to higher carbon stock in the plantation forest in the future.

Table 2: Above Ground Biomass and Carbon Content in Different Forest Manipur, North East India

Sl.No.	Forest Type	Altitude (m)	Age	Biomass (t ha^{-1})	Carbon Content (t ha^{-1})	Location	Sources of Data (Ref.)
1.	Oak forest						
	Site-I	1800	30	186.96	93.48	Shiroy hill,	[8]
	Site-II			211.50	105.75	(Manipur)	
2.	Oak forest						
	Site-I	820	40	202.74	101.37	Maram	[6]
	Site-II		38	192.18	96.09	Manipur	
	Site-III		32	130.22	65.11		
	Site-IV		42	255.04	127.52		
3.	Pine Plantation	810					
	forest		16	173.47	86.73	Imphal	[7]
			20	220.00	110.00	valley	
			24	274.00	137.00		
			30	357.00	178.50		
			40	591.00	295.50		
4.	Secondary *Dipterocarpus* Forest	330					
	Sites-I		16	21.92	10.96	Moreh,	[10]
	Sites-II		16	18.78	9.89	Manipur	

RATE OF CARBON SEQUESTRATION

The rate of carbon sequestration in different forest ecosystems is set in Table 3. The rate of carbon sequestration was highest in oak forest ranging from 11.89 to 15.90 t C ha^{-1} and lowest in *Dipterocarpus* forest ranged from 3.21 to 3.77 t C ha^{-1}. In pine forest rate of carbon sequestration varied from 6.24 (16 years old) to 13.72 t C ha^{-1} (40 yrs old).

The rate of carbon sequestration is comparatively higher than other values reported for Philippines forest ecosystem, but lower than the value reported by Ryan *et al.* [14] in *Pinus rediata* plantation in Canberra, Australia and *Pinus elliottii* plantation in Florida, USA [15]. The rate of sequestration in old growth tropical forest of Brazil is higher (6.50 t C ha^{-1}[16] than tropical *Dipterocarpus* forest of Manipur, Northeast India. The rate of carbon sequestration is highly variable and depends on the species composition, nutrient status of soil, age of tree and other climatic factors.

SOIL CARBON

Soils store different amount of organic carbon depending on the amount and quality of plant input material *i.e.* above ground and below ground litters, root exudates and their turnover.

Table 3: Rate of Carbon Sequestration in Different Forests of Manipur, NE India

Sl.No.	Forest Type	Altitude (m)	Age	Net Biomass Production (t $ha^{-1}yr^{-1}$)	Carbon Flux C ($ha^{-1}yr^{-1}$)	Location	Sources of Data (Ref)
1.	Oak forest						
	Site-I	1600	30	24.22	12.11	Shiroy hill,	[8]
	Site-II		30	23.74	11.89	Manipur	
2.	Oak forest	820					
	Site-I		40	27.37	13.69	Maram	[6]
	Site-II		38	29.05	14.52	Manipur	
	Site-III		32	28.39	14.20		
	Site-IV		42	31.81	15.90		
3.	Pine Plantation	810					
	forest		16	12.48	6.24	Imphal	[7]
			20	14.26	7.13	valley	
			24	16.98	8.49		
			30	17.48	8.74		
			40	27.44	13.72		
4.	Secondary *Dipterocarpus* forest	330					
	Sites-I			7.54	3.77	Moreh,	[10]
	Sites-II			6.43	3.21	Manipur	

The data on soil carbon is limited. The soil organic matter in the soils in the upper layer (0-30 cm) ranged from 27.73 (Oak forest) to 48.03 t C ha^{-1} (*Dipterocarpus*) Table 4. Higher value in tropical *Dipterocapus* forest may be due to high rate of litter production and faster decomposition of litter [17]. Most of the soil carbon data are based on 30 cm soil depths which may account for only a small fraction of soil carbon. Percentage of the total carbon in soil is highly variable and ranged from 6.80 to 70.24 per cent out of the total carbon pool in vegetation and soil (Table 5). Thus the forest soils may contain 2 to 3 times more soil carbon as compared to reported value on the basis of top 30 cm soil. The information on the soil carbon distributed in the deeper layers in roots is lacking which hold a great potential of carbon sequestration in the soil. Therefore the role of soil carbon in the roots and soil carbon leaching in deeper layers needs to be investigated.

EMISSION OF CARBON DIOXIDE FROM SOIL

The annual rate of emission of CO_2 from the soils under different forest ecosystems of Manipur varied from 1.697 to 4.462 t CO_2-C $ha^{-1}yr^{-1}$ from the soils in Oak and *Dipterocarpus* forest (Table 6). Tropical forest of *Dipterocarpus* exhibited high rate of CO_2 emission than that of Oak forest even though with low carbon stock and rate of sequestration in the former. It seems that CO_2 emission from the soils does not depend only on carbon stock and rate of carbon sequestration in vegetation, but also on abiotic variables.

Table 4: Mean Organic Carbon Content in the Soils of Different Forests of Manipur, NE India

Sl.No.	Forest Types	Altitude (m)	t C ha^{-1}	Location	Sources
1.	Oak forest	820	27.73	Maram, Manipur	[6]
2.	Pine plantation	810	22.56	Imphal Valley	[7]
3.	*Diptrocarpus* forest	330	48.03	Moreh, Manipur	[9]

Table 5: Carbon Stock in Vegetation and Soils (t C ha^{-1}) in Different Forests of Manipur, NE India

Forest Type	Carbon Stock in Vegetation	Carbon Stock in Soil	Per cent of Total Carbon in Soil	Sources
Oak forest	195.45	27.73	22.51	[6]
Pine forest	319.09	22.56	6.80	[7]
Dipterocarpus forest	20.35	48.03	70.24	[17]

Table 6: Annual CO_2 Emission from the Soils in Different Forests in Manipur, NE India

Sl.No.	Forest Types	t CO_2 $ha^{-1}yr^{-1}$	Sources of Data
1.	Oak forest	4.462	[18]
2.	Sub-tropical mixed oak forest		
	Stand-I	1.697	[10]
	Stand-II	1.478	
3.	*Dipterocarpus* forest		
	Control site	3.500	[9]
	Logged site	3.332	
	Burnt site	4.0482	

High values of CO_2 emission mainly depend upon organic matter and climatic factors *i.e.*, temperature and soil moisture [10,18].

Several studies reported that temperature was the single most important variable for predicting the soil CO_2 flux [19, 20]. According to Chapman and Thurlow [21] rise in the mean annual temperature of 5°C could potentially increase CO_2 emission by a factor of 2 to 4. CO_2 emission is strongly influenced by seasons and abiotic variables as reported by Devi and Singh [10] in sub-tropical forests.

Kyoto protocol provides a flexible mechanism to meet carbon reduction obligation and Clean Development Mechanism (CDM) permits the countries to meet their carbon reduction quota in developing countries. Therefore, tropical counties may actively engage in mitigation CO_2 through mass scale reforestation and afforestation programme.

CONCLUSIONS

Forests exhibit wide range of carbon stocks in the State of Manipur. Pine forest exhibits highest carbon stock 295.00 t C ha^{-1} followed by oak forest (65.11 to127.52 t C ha^{-1}) and *Dipterocarpus* forest (9.89 to 10.96 t C ha^{-1}). The rate of carbon sequestration is maximum in oak forest (11.89 to 15.90 t C ha^{-1}) and minimum in secondary *Dipterocarpus* forest (3.21 to 3.77 t C ha^{-1}).

Soil organic carbon ranged from 27.73 to 48.03 t C ha^{-1} which is highly underestimated. Emission of CO_2 varies from 1.697 to 4.462 t CO_2 in different forest soils. It seems that carbon stock, rate of sequestration and emission of carbon dioxide in the different ecosystems is governed by species composition, age of trees and climatic factors.

However, further research on carbon stock and carbon sequestration in soil and vegetation is urgently needed for sustainable development of forest resources and to mitigate climate change in the long term. It will also provide the basic information to assess the carbon trading options and the societal value of carbon.

REFERENCES

[1] Brown, S., J. Sathaye, M. Cannel and P. Kauppi, 1996, Management of forests for mitigation of greenhouse gas emission, Chapter 24. In: Climate change 1995: Impact Adaptation and mitigation of climate change: scientific-technical analysis (RT Watson, MC Zinyowera and RH Moss, Eds.) Contribution of working group II to the second Assessment Report of the Inter-government Panel on Climate change, Cambridge University Press, Pp. 775-797.

[2] Dadhawal.V.K. and A. Shah, 1997, Recent changes in forest phytomass carbon pool India estimated using growing stock and remote sensing based on forest inventories. Journal of Tropical forestry, *13*:182-188.

[3] Ravindranath N.H., B.S. Somashekhar and Madhav Gadgil, 1997, Carbon flows in Indian forests climate changes, Climate Change, 35:297-320.

[4] Chatterjee, S.P., 1965, Physiography, The Gazette of India, Vol. I, Ministry of Information and Broadcasting, Government of India, Delhi.

[5] Census of India, 2001, Directorate of Census Operation, Manipur.

[6] Srinivas, C., 1992, Plant biomass, net primary productivity and nutrient cycling in oak forests of Manipur, Ph. D Thesis, Manipur University, Imphal, India.

[7] Devi, K.B., 1993, Primary production and nutrient cycling in *Pinus kesiya* forests of Manipur. Ph. D. Thesis. Manipur University, Imphal, India.

[8] Singh, E.J and P.S., Yadava, 1994, Structure and function of oak forest ecosystem of North East India, 1: Biomass dynamics and net primary production, Oecologia Montana, 3: 1-9.

[9] Yadava P.S., 2007, Impact of human activities on the soil microbial biomass and nutrient mineralization in the *Dipterocarpus* forest of Manipur, N.E. India. CSIR Annual Report, Manipur University, Imphal, India.

[10] Devi, N. B. and P.S. Yadava, 2009, Emission of CO_2 from the soil and immobilization of carbon in microbes in a sub-tropical mixed oak forest ecosystems, Manipur, N.E. India, Current Science, 96(12):1637-1630.

[11] Houghton, R.A., 2000, Interannual variability in the global carbon cycle, J. of Geophys. Res., 105: 20121-20130.

[12] Lasco, R.D. and F.B. Pulhin, 2003, Philippine forest ecosystem and climate changes: Carbon stock, rate of sequestration and Kyoto protocol, Annals of Tropical Research, 25(2): 37-51.

[13] Devi, L.S. and P.S. Yadava, 2009, Aboveground biomass and net primary production and semi-evergreen tropical forest of Manipur north-eastern India. Journal of Forestry Research, 20(2):151-155.

[14] Ryan, M.G., R M Hubbard, S. Pongracic, J.R, raison, R E. McMurtre, 1996, Foliage, fine roots, wood tissue and stand regeneration in Pinus radiata in relation to nutrient status, Tree Physiology, 16: 333-343.

[15] Gholz, H L and R.F. Fischer, 1982, Organic matter production and distribution in slash pine (Pinus elliottii) plantation. Ecology, 63:1027-1839.

[16] Malhi, Y, D.D. Baldocchi and PG. Jarvis, 1999, The carbon balance of tropical, temperate and boreal forests, Plant, Cell and Environment. 22:715-740

[17] Devi, A. Surjubala and P.S. Yadava, 2007, Wood and leaf decomposition of *Dipterocarpus tuberculatus* Roxb. ecosystem at Shiroy hills, Manipur, Northeastern India, International Journal of Ecology and Environmental Sciences, 28:133-137.

[18] Laishram, I.D., P.S. Yadava and L.N. Kakati, 2002, Soil respiration in a mixed oak forest ecosystems at Shiroy hills, Manipur, Northeastern India, International Journal of Ecology and Environmental Sciences, 28:133-137.

[19] Bijracharya, R.M., R. Lal and J.M. Kimbe, 2000, Durinal and seasonal CO_2 C flux from soil as related to erosion phases in Central Ohio., Soil Sci. Soc. Am. J., 64:286-293.

[20] Rastogi, M., S. Singh and H. Pathak, 2002, Emission of carbon dioxide from the soil, Current Science, 82:510-517.

[21] Chapman, S.J. and M. Thurlow, 1998, Peat respiration at low temperature, Soil Biol. Biochem., 30:1013-1021.

Chapter 15

CO_2 Mitigation: Issues and Strategies

V.S. Verma*

Member, Central Electricity Regulatory Commission and Former Member (Planning), CEA

SUMMARY

The Indian Power Sector is currently growing at a rapid pace. Major capacity addition programe for the current 11th Plan (2007-2012) and 12th & 13th Plans (2013-2022) is based on supercritical technology. In this paper, growth of electricity scenario, various energy efficiency improvement measures and issues in carbon capture and storage are discussed. CCS is not the appropriate option for India, but India should pursue its own R&D in the area of carbon capture and fixation. Pros and cons of adopting CO_2 mitigation measures other than CCS are discussed. Total CO_2 emissions and average emission factors with Renewable Energy Sources (RES) has been computed. Regulatory and technology measures for reducing CO2 emissions further also are described.

INDIAN POWER SECTOR

The total installed power generation capacity in the country is over 1,55,700 MW consisting of 99,600 MW thermal (64 per cent), 36,800 MW hydro (24 per cent), 4,100 MW nuclear (2.5 per cent) and about 15,200 MW renewable (9.5 per cent). The thermal capacity comprises of 81,600 MW coal and lignite fired, 16,800 MW gas fired and diesel 1200 MW. We generate about 758 billion units per year. Similarly, the peak

Guest lecture delivered in the Awareness and Capacity Building in Carbon Capture and Storage Programme (ACBCCS-2009) conducted from July 27-31, 2009 at Indian National Science Academy, Delhi.

generation in the country is around 100,000 MW. We are short of 13 per cent during the peak and about 10 per cent in energy. Most of the generation (about 69 per cent) comes from coal fired power plants, 15 per cent from hydro, 10 per cent from gas fired and diesel, 2 per cent from nuclear and less than about 4 per cent from renewables.

While planning for electricity in the country from different sources it is seen that there are specific constraints with each type of sources–be it hydro, which have long gestation period and also these are located in difficult terrains; or gas, there are limitations of availability of gas in the country and also its uncertain prices. With renewable, the cost of generation is very high. The wind and solar taken together would generate about 20 per cent of total renewables. The wind and solar generation is also not despatchable electricity because of uncertainty of their availability. There is uncertainty of the availability of nuclear fuel as well. Accordingly, the major share of generation goes to coal fired power plants. Coal is indigenously available in sufficient quantity perhaps for next 60-70 years in a *business-as-usual* scenario. Nevertheless, coal would continue to be the main source of power generation in the country.

Future Plan Projections

It is pertinent to mention that the country is planning to add a capacity of about 80,000 MW in the 11th Plan and about a lakh MW each in 12th and 13th Plans, the major share of which would go to the coal.

The present level of energy efficiency of generation in the coal fired power plants in the country is about 34 per cent which is proposed to be raised to a level of 36 per cent by the end of 11 Plan and 38 per cent by the end of 12th Plan. With the addition of the state-of-the-art technology like super critical as well as other measures including retirement of old and inefficient units of less than 200 MW in size, and through Renovation and Modernization (R&M) of the old coal fired units, hydro and nuclear power units to maintain their availability and efficiency at reasonably high level. Priority is also being accorded for gas based and hydro power generation.

The coal fired power plants are the main source of CO_2 emission. The power sector alone emits about 520 million CO_2 in absolute terms (Table 1).

Table 1: Year-wise Absolute Carbon Dioxide Emissions (in million tons)

	2002-03	*2003-04*	*2004-05*	*2005-06*	*2006-07*	*2007-08*
North	106.8	109.9	112.2	120.1	129.5	406.56*
East	66.5	75.5	83.9	92.5	96.3	
West	148.5	144.1	157.7	153.9	157.7	
North-East	2.2	2.4	2.4	2.5	2.6	
South	105.3	108.2	105.6	101.1	109.2	113.62
India	429.4	440.2	462.02	470.8	495.53	520.18

*: Northern, Eastern, North Eastern and Western Regions have been integrated as one grid from year 2007-08.

The source of all factual technical information/statistics is CEA.

The specific CO_2 emission in the country from coal based power plant is 1.08 tons of CO_2/MWh based on Net generation. Weighted average specific emissions for fossil fuel based power plants in 2007 are shown in Table 2. If super-critical technology is deployed the CO_2 emission from coal fired stations could be brought down to 0.94 tons of CO_2/MWh. The average CO_2 emission per unit of electricity generation (net) in the country has been brought down to about 0.8 tons of CO_2 in a period of about 4-5 years.

Table 2: Weighted Average Specific Emissions for Fossil Fuel Fired Stations in FY 2007-08 (Figures in CO_2/MWh)*

Coal	*Diesel*	*Gas*	*Lignite*	*Naphtha*	*Oil*
1.08	0.62	0.47	1.40	0.45	0.74

* Based on Net generation.

Government has already notified the National Action Plan for Climate Change, a very comprehensive document enumerating about the policies to reduce the carbon footprint of the country especially in the power sector. The various provisions of the National Action Plan are in various stages of implementation through missions being set up by the Government, like the Solar Energy Mission and others. The subject of mitigation of CO_2 in the country has acquired significance of an economic issue rather than merely CO_2 mitigation and has to be dealt with in totality reference to our requirements and priorities as well as economic feasibility.

The pros and cons of adopting various measures in this direction are discussed in the following paragraphs.

ISSUES IN CARBON CAPTURE AND STORAGE

Since it is imminent for us to burn coal for power generation as a major share, the carbon capture and storage could be one of the alternatives to mitigate the emission of carbon dioxide. This would involve adoption of a technology for carbon sequestration, *i.e.* separation of CO_2 from the flue gases and taking the same to the remote geological spaces for safe storage. Alternatively the CO_2 could also be utilized for enhanced oil and gas recoveries. As per the present development elsewhere in the world, the CCS technology has not been proven on commercial scale, although efforts are on in the direction of the development of pilot projects. Some of the manufacturers are monopolizing the design, manufacture and supply of plant and equipments required for such plants. The costs are also on higher side from our perspective. The information available in this regard through various publication/literature/internet and interactions with the international manufacturers, utilities indicate that capital cost of plants with such installations would be about 200 per cent of the plant and equipment for conventional coal fired generation. This would result into almost doubling the cost of generation of electricity in the country. Further, the auxiliary power consumption requirement of these plants would be about 40 per cent of the total generation from the plant, accounting for reduction of efficiency of generation from the coal fired power plant by about 12-15 percentage. Coal consumption would also increase by about 30 per cent.

The first principles of the design and engineering of this technology is yet to be firmed up and is very uncertain. There are issues of technology transfer and funding of the same as far as our country is concerned. As per some of the technology papers presented through UNFCCC discussion/conferences there appears to be a lack of consensus on the adoption of these technologies for the time being. In addition, there are important issues like safety, security and disaster management in case there is any leakage or other calamities which are to be addressed appropriately before adoption. Of course, there are issues of priorities, availability of electricity and other social requirements in our case. However, there is a need for extensive research and development activity indigenously with focus on various intricate issues and areas of concern for us specifically.

We are looking for some alternatives to the above proposal to suit our requirements. One such idea is carrying out research and development at the first instance with indigenous efforts in the area of carbon sequestration and fixation. It is understood that few laboratories in the country are already on the job. Some funds would need to be set aside to carry out research in this area so that the outcome of the research would address the whole issue of carbon capture and fixation, etc. rather than storage. Even if financial assistance is forthcoming from the developed world to set up pilot projects in the country, this is fraught with the uncertain future for installations on other units because of constraints arising out of the matters discussed in the preceding paragraphs.

OTHER OPTIONS AND STRATEGIES

Other options and strategies to improve energy efficiency and CO_2 reduction are as follows.

Increasing Efficiency of Generation of the Existing Power Plants

It is rather a very desirable exercise to modernize the existing power generating units which results in higher efficiency of generation through Renovation and Modernization and which would also result into substantial reduction of CO_2 emissions. This is indicated as an example in the Table 3, considering typical similar units of 210 MW capacity with conservative numbers.

It can be seen that one 210 MW unit can result in savings of CO_2 emission of about 290,000 tons per annum. This is very substantial reduction. The country has the operating units in various sizes which are indicated in the Table 4. A capacity of about 20,000 MW has been proposed to be taken up for renovation and modernization. The reduction of CO_2 emission would thus be enormous.

Retirement of Old and Inefficient Units

Retirement and replacement of old and inefficient plants and replacing with more efficient ones is an effective way of reducing fuel consumption and minimizing the CO_2 emission. As per the indications available from CEA, about 1500 MW capacity is to be retired in 11th Plan. These units would consists of sizes less than 100 MW and few 110 MW units, all gas based units more than 30 years old to be retired, about 7000 MW capacity are stated to be retired during 12th Plan, all coal based units of 200 MW capacity and commissioned prior to the year 1982 totaling to 3,700 MW are to be retired during 13th Plan, gas units commissioned prior to the year 1992 would also to be retired during 13th Plan. The total capacity of about 12,200 MW would see the retirement by the end of 13th Plan.

Table 3: CO_2 Emission Reduction Before and After R&M

Parameter Unit	*Before R&M*	*After R&M*
Capacity MW	210	210
Plant Load Factor per cent	75	75
Heat Rate kcal/kWh	2860	2280
Net efficiency per cent	30	37.7
Annual Generation MU	1379	1379
Specific Coal Consumption kg/kWh	0.71	0.57
GCV in kcal/kg	4000	4000
CO_2 emission kg/kWh	1.04	0.83
Total CO_2 Emission in 000' tons	1434	1145
Reduction in CO_2 emission in 000' tons	289	

Table 4: Various Unit Sizes and Main Parameters

Unit Size MW	*MS Pressure kg/cm²*	*MS/RH Temperature ºC*	*Gross Design Efficiency (per cent)*
30-50	60	482	28.20
60-100	90	535	31.30
210 Z	130	535/535	35.63
210 U	150	535/535	37.04
250	150	535/535	38.3
500	169	538/538	38.6*

The above action would result in raising the average efficiency of generation by about 3 4 per cent. The CO_2 reduction against this would be substantial.

Reduction of T&D Loss

The all India T&D loss which are around 28 per cent are aimed to bring down to about 15 per cent by the year 2011-12. This would again result in saving of power consumption and consequent reduction in CO_2 emission.

Improvement in the Coal Quality by Blending, Washing, etc.

This option is also being considered for implementation and would result into CO_2 reduction due to the positive effects on Boiler efficiency. A policy of washing of coal for thermal power generation would be desirable.

Monitoring on Regular Basis the Availability and Efficiency of Power Plants

To sustain higher efficiency of generation in the country, various actions through mapping of thermal power stations, devising of standard energy audit methodology for accredited energy audit of the power plants, the establishment of energy efficiency cells itself in the power station etc. are being taken up. These are being actively and aggressively being pursued by the government with the concerned utilities. The Excellence Enhancement Center is being set up in the country to keep a watch and render expert advice on tackling various issues relating to power generation, transmission and distribution. This is being achieved through various international cooperation like the Indo-German Energy Programme.

Adoption of Advanced Coal Technologies in Thermal Generation Units

Adoption of super critical technology gives about 2 per cent gain over sub critical technology consequently resulting into reduction of CO_2 emission. The initial super critical units in the country are based on steam parameters of 246 kg/cm^2 pressure and 535/565°C temperature at turbine inlet. Higher steam temperature of 565/593°C are also being adopted at newer stations which would result into further gains of efficiency in generation. The efficiency gains of various super critical parameters are indicated in the Table 5.

Table 5: Super Critical Coal Combustion Parameters and the Expected Efficiency Gain

MS Pressure bars	*MS/RH Temperature ºC*	*Gross Efficiency per cent*	*Efficiency Gain*	*Cumulative Efficiency Gain per cent*	*Relative Efficiency Gain per cent*	*Remarks*
169	538/538	38.6	Base	Base	Base	Present 500 MW units
246	538/538	39.29	0.69	0.69	1.8	Super critical Units
246	538/566	39.56	0.27	0.96	2.49	Super critical Units
246	566/566	39.91	0.35	1.31	3.39	Super critical Units
246	566/593	40.24	0.33	1.64	4.25	Super critical Units
246	600/600	40.56	0.32	1.96	5.08	Super critical Units

The Government is according very high priority on adoption of super critical technology in the country for future power generating plants. This can be seen in Table 6 given below:

Table 6: Plan Targets for Addition of Supercritical Generation

Five Year Plan	*No. of Units*	*Capacity (MW)*
11th Plan	14	9,520
12th Plan	76	52,260

In the 13th Plan, thermal capacity will be added by projects based on super critical technology only (please see box).

Capacity Addition Programme Based on Super Critical Technology

- 11 number super critical units totaling to 7,540 MW under construction for likely benefits during 11th Plan.
- 20 number super critical units totaling to 14,000 MW ordered so far for likely benefits during 12th Plan.
- 13 number ultra mega power units, each of about 4,000 MW capacity, based on supercritical technology planned. Four units already awarded.
- About 62 number super critical units totaling to 44,000 MW likely to be added during 12th Plan.
- Coal based capacity likely to be added during 13th Plan and beyond to be almost entirely super critical.

After prolonged deliberations with the Executive Board of the CDM it has been agreed to consider the super critical technology of coal fired power plants for the CDM benefits. But the various conditionalities imposed result into the gain in the CDM benefits practically only for the initial installations. The moment any super critical plant get commissioned, its gain becomes a part and parcel of baseline data and in case more than 3-4 units gets commissioned the gain practically will not be there unless we set up ultra super critical plants with very high steam temperature and pressures. There is a need to keep in touch with the Research and Development of steam generating units with 700°C plus temperature at the turbine inlet being carried out by the European Countries as a collaborative attempt. If succeeded, this would be a big milestone in the field of reduction of carbon footprint in the conventional coal fired generation.

IGCC Technology

The integrated gasification combined cycle technology (IGCC) refers to power generation through gasification of coal by the combined cycle power plant. The Indian coal having high ash content do not support itself to easy gasification process especially when it comes to acceptable quality of gas for firing into the gas turbine. The cleaning of the gases itself becomes a bottleneck and big hurdle for achieving higher efficiency on this technology with the Indian coal. Government is encouraging pilot research projects on IGCC since this technology has the potential to achieve higher efficiency of generation and consequent lower CO_2 emission. This is seen as a future promising technology in the direction of low carbon growth in the power sector.

Encouraging the Gas Based Co-generation System

The efficiency of combined heating, cooling and power production is very high to the tune of 65 per cent. This option also results into lesser land and water requirement, almost negligible T&D losses and increased reliability of service. This is a very desirable option to be adapted wherever feasible.

Initiative Under National Action Plan on Climate Change

Eight missions have been set up under the National Action Plan on Climate Change. One such mission relates to power sector called National Mission for Enhanced Energy Efficiency. The scheme would incentivise increase in energy efficiency in energy intensive industries in India. Trading in energy efficiency is also being encouraged.

Use of Renewable Source of Energy

Any generation from the renewable source of energy is a direct gain in CO_2 reduction and high priority is being accorded to this area. The solar energy mission has been set up with high targets of capacity addition from the solar thermal plants. It is understood that the NTPC has taken the first initiative to set up 1000 MW capacity of solar installations in the country which would see the light of the day during the 11th Plan itself. There are issues of cost of generation, etc. on the renewable source of energy which are to be tackled from the point of view of their easy adoption. There is an urgent need to initiate and sponsor research in the area for bringing down the cost of plant and equipment specially in the solar area so that the technology become affordable. It is to be understood that any technology which requires subsidies would not be a sustainable option on a long term basis. Net generation emission factors and CO_2 emissions (with renewable) at the end of 11th, 12th and 13th plan is shown in Table 7. Average emissions factor on total generation with RES is shown in Table 8. Total CO_2 emissions at the end of plan period with renewables and with RES are shown in Figure 1 and Figure 2.

Table 7: Emission Factor on Thermal Generation and CO_2 Emission at the End of 10th, 11th, 12th and 13th Plans (With Renewables) Base Case

Period	*Emission Factor on Gross Thermal Generation kg/kWh*	*Net Thermal Generation (Coal + Lignite + Gas–Renewables) MU*	*Total CO_2 Emission (With Renewables) Million Tons*
At the end of 10th Plan	0.944	517104	488
Anticipated at the end of 11th Plan	0.868	802728	697
Anticipated at the end of 12th Plan	0.842	1011814	852
Anticipated at the end of 13th Plan	0.819	1330040	1081

Figure 3 shows the total CO_2 emission reduction with RES.

The Central Electricity Regulatory Commission has recently brought out regulations on the renewable source of energy bringing out norms of capital cost and

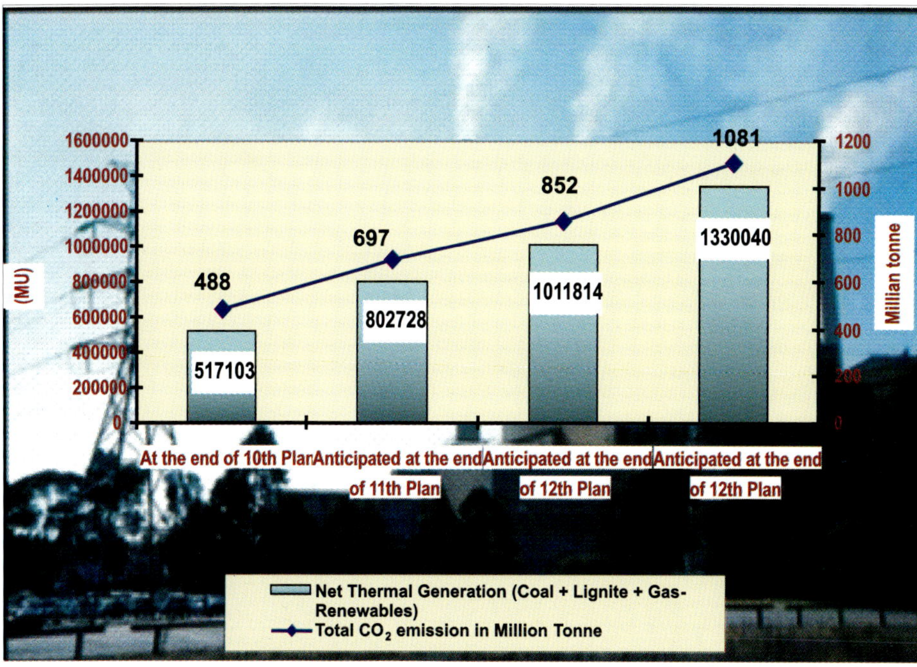

Figure 1: Total CO_2 Emission at the End of 10th, 11th, 12th and 13th Plans

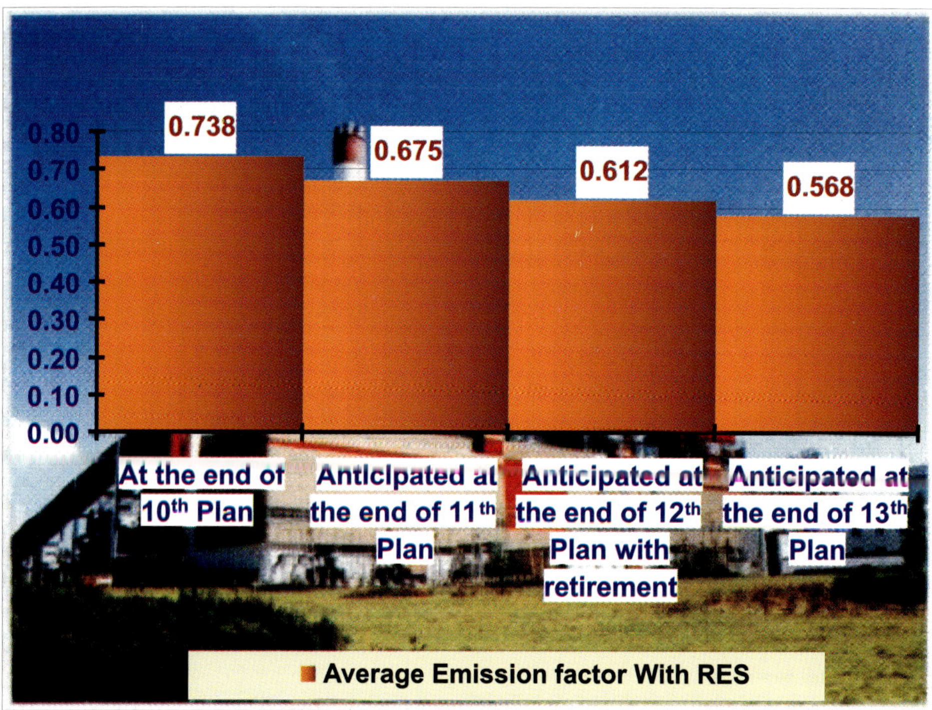

Figure 2: Average Emission Factors With RES

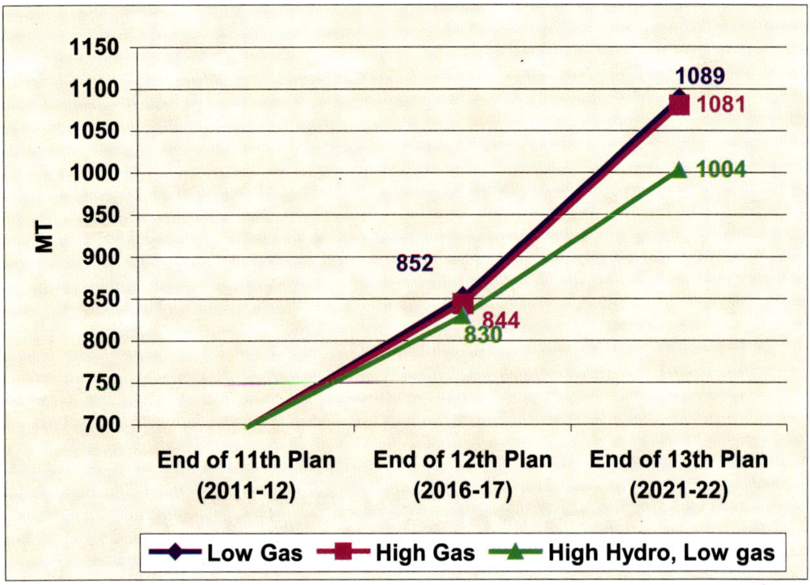

Figure 3: Total CO_2 Emission (MT) with RES

other financial parameters as well as the normative tariff to be adopted for such technologies. These are very encouraging and are proposed to be reviewed after every one year/3 years to update the capital cost as well as other financial norms. Once the power plant is established, the given tariff would be applicable for the life of power plant. This would increase the share of renewable source of energy and ultimately bring down the CO_2 emission.

Table 8: Average Emission Factor on Total Generation (With RES) at the End of 10th, 11th, 12th and 13th Plans (Base Case)

Period	*Total Generation (MU)*	*Average Emission Factor with RES (kg/kWh)*	*Thermal Generation (per cent)*
At the end of 10th Plan	661277	0.738	78.20
Anticipated at the end of 11th Plan	1033040	0.675	77.71
Anticipated at the end of 12th Plan with retirement	1393533	0.612	72.61
Anticipated at the end of 13th Plan with retirement	1915854	0.568	69.42

RESEARCH AND DEVELOPMENT

The following are a few suggested areas of research and development which needs to be taken up/sponsored on urgent basis to deal with climate change on priority.

1. Carbon sequestration and fixation
2. Further increase of efficiency of conventional coal-fired power plants. This would entail the development of suitable materials for application for high temperature
3. Development of clean coal technologies like IGCC, PFBC, coal bed methane, ultra super-critical technologies, etc.

Note: The views expressed are not necessarily those of the Organisations to which the author belong.

Chapter 16
Should Carbon Be Priced?

Open Round Table Discussion

The highlights of the view expressed by the Panelists are as follows:

Shri V.S. Verma

Member, CERC chaired the Open Round Table Discussion.

Sh. V.S Verma in his opening remarks said that his views on financing of CCS projects are already expressed in the previous chapter on CO_2 Mitigation: Issues & strategies. There appears to be no consensus on the adoption of carbon capture and storage technologies for the time being under the Clean Development Mechanism (CDM) as seen from the technology documents presented through UNFCCC discussion/conferences. Suggestions have also been made by some countries to include CCS in CDM. In this context, after a prolonged deliberations with the Executive Board of the CDM it has been agreed to consider the super-critical technology of coal fired power plants for the CDM benefits. But the various conditionality imposed resulted into the gain in the CDM benefits practically only for the initial installations. The moment any super critical plant get commissioned, its gain becomes a part and parcel of baseline data and in case more than 3-4 units gets commissioned the gain practically will not be there unless we set up ultra super-critical plants with very high steam temperature and pressures.

Dr. (Mrs) Malti Goel

Former Adviser, Ministry of Science and Technology

One of the critical issues about the theme of the discussion, Should Carbon be priced? is, should CCS be included in Clean Development Mechanism (CDM)? The CDM is a market based mechanism for clean development, which came into being in 2005 after Kyoto Protocol has been ratified under UN Framework Convention on Climate Change (please see box). CDM is designed to support sustainable

CDM: A KYOTO PROTOCOL MECHANISM TO PROMOTE CLEAN DEVELOPMENT

The clean development mechanism (CDM) as an idea was proposed under Article 12 of Kyoto Protocol with the objective of transfer of energy efficient technology from Annex-I to non Annex-I countries. Under this mechanism Annex–I countries would invest in plant and equipment in participating non Annex-I countries and get the carbon credit against 2008-12 target of reducing GHG emissions of Annex-I countries. The CDM was designed in such a manner, so as to lower the overall cost of reducing GHG emissions and support sustainable development in non Annex-I countries.

Baseline for CDM

Any entity (foreign, domestic, joint venture, public, corporate, non-profit) may set up a project to reduce CO_2 emissions in a non Annex-I country. The CDM Executive Board decides on existing actual or historic emissions as Baseline Data. Determination of baseline data requires

- ✰ Calculation of GHG emissions by sources
- ✰ Emissions from technology attractive investment
- ✰ Average emissions of similar technology in previous 5 years performance
- ✰ System boundaries and leakages

Certified Emissions Reductions [CERs] are issued by CDM Executive Board, which can be transferred to the buyer in Annex-I country. Both Annex I and non Annex-I Parties should have ratified Kyoto Protocol. It also allows trading of CERs in global market such as Emission Trading Scheme (ETS).

Perogative of the Host Country

The CDM allows host country approval for emissions additionality criteria, sustainable development goals and seeks voluntary participation on account of their economic well being, social well being and environmental well being.

Approval of CDM Executive Board

The Board has registered total of 1650 projects with approved methodologies registered so far from different countries adding to about $28 billion in 2009. Meth Panel set up by CDM Executive Board determines the scope of approved methodologies. In CDM the market price of CO_2 has been varying from time to time.

India and CDM

India has done well by capturing major share in CDM approved projects. Indian industry has come forward to actively participate in this marked based mechanism. More than 700 companies have proposed projects to National CDM Authority in Ministry of Environment and Forests. So far around 1100 projects have been sent from India to CDM Executive Board by the National CDM Authority, out of this one third share of projects are already registered by CDM. These projects are mainly in the area of; Renewable energy (Wind, Biomass, Solar, Hydro), Switching to Alternate Fuels, Co-generation, Energy Efficiency, Waste Management, Oil sector, Agriculture and Forestation and others.

development in non-Annex I countries. India has fared well by capturing major share in CDM approved projects but the issue on CCS is still debated and inclusion of clean coal technology has not been encouraged.

She said other issues for discussion could be:

- ✰ How CCS projects are getting investment and how to address disputes in baseline determination of certain technologies?
- ✰ If CCS is included in CDM should CO_2 pricing in Certifies Emission Reductions (CERs) for CCS be made dependent on relative cost of technology and risk involved?
- ✰ What market design changes and other trading schemes are evolving?
- ✰ What should be the approach in terms of technical issues such as, greenhouse gas emission controversies, support to R&D, technology transfer, etc.?

Pamposh Bhat

Pamposh Bhat, Expert Climate Change, CEO, EFCON

- ✰ Globally, the carbon trading market was worth •92bn (£79bn) in 2008, trading 5b tones of carbon dioxide under Emission Trading Scheme (ETS). The ETS covers 50% of the UK and EU's carbon emissions, mainly in the energy, cement, steel, glass and manufacturing sectors. Companies in these sectors have allocated allowances for the carbon they emit, with the total number shrinking over time, theoretically forcing companies to buy additional permits to pollute if they do not cut emissions. Large proportion of the UK's promised cut of 34% by 2020 will come via British companies in the ETS.
- ✰ However, the large number of carbon permits that have been allocated and a fall in emissions due to the recession have made the trading system less effective.
- ✰ "With too many rights to pollute in circulation, the scheme is in danger of being rendered irrelevant," said Sandbag founder, Bryony Worthington. "At a time when other countries are looking to set up their own trading schemes and the world is set to debate a global deal on how to tackle climate change, flagship policy urgently needs rescuing – starting with much tougher caps."He called for an immediate tightening of the cap on permits to 30 per cent of industry's emissions by 2020, compared to the existing 21 per cent, and a commitment to 40 per cent if a strong global deal results from a UN climate change summit in Copenhagen in December. Making the 30 per cent cut would cost virtually the same as was originally envisaged for the 21 per cent cut, she said, and be much closer to the cuts scientists say must be made to avoid dangerous climate change. But MP Tim Yeo, chair of the environment audit committee, said: "These findings confirm what many have begun to suspect. Although emissions trading

remain conceptually valid, in practice the EU ETS has not succeeded in driving investment in low-carbon technology."

- It is said that the ETS price for a tonne of CO_2 is around €14. To make it economical for generators to switch from coal to less-polluting fuel like Natural gas for electricity production requires a price of around €25, while carbon capture and storage technology would need a price of €40-€50 a ton to be worth investing in.
- The companies have the option to offset their emissions by buying credits from outside the EU, usually from hydroelectric or other schemes in China and India. On current trends, 900million of these could be available up to 2012, and bankable for use up to 2020.
- The non-EU credits come mainly from the UN's Clean Development Mechanism, which is widely acknowledged to be flawed. It includes many projects that would have happened without CDM funding, meaning the carbon reductions are not true cuts. Campaigners also argue it allows rich nations a "get out of jail free card", when they should be making cuts in their own countries.
- ETS was not being effective in tackling global warming. As the most mature trading system, it is seen as a model for newer markets around the world, which will need to be integrated for a truly effective global system of cutting emissions. But R.K. Pachauri, head of the Intergovernmental Panel on Climate Change, said a strong Copenhagen agreement could lead to a substantial shift in the carbon market, lifting the price, "It may change the whole dynamic. That is my feeling, though I may be wrong". Markets such as the Chicago Climate Exchange now allow companies to voluntarily limit their carbon emissions and lower their carbon footprint by purchasing credits, traded on the market like stock.
- While opinions differ several studies have shown that CCS is technologically feasible and a good option to reduce carbon emissions from power plants. Technology exists now — and is being used — to capture carbon dioxide from power plants, and then transport it as a liquid and store it underground.
- A study conducted by the Electric Power Research Institute in the US argued that it is possible to reduce — through CCS — about 350 million tonnes of carbon dioxide per year in the US alone by 2030.
- Most of the work is concentrated on the large power plants, which contribute 40 per cent of the carbon dioxide emissions. But, for all these processes to work in the long-term, one might need to find a way to convert the captured carbon dioxide into a useful commercial product, and not just bury it somewhere. This is another line of research, and an entirely different story.

Dr. Nittala S. Sarma

Professor, Andhra University, Visakhapatnam

Prof. Sarma observed that to the 120 terragrams (Tg) of carbon injected into the atmosphere during the last 50 years, an overwhelming contribution is from the developed countries. Without fixing responsibility for this happening on those countries, it would amount to a partisan treatment if now developing countries are asked to limit their carbon emissions, for the vehicle for development is essentially a carbon-intensive one. But unfortunately, the advanced countries are unprepared to accept their role in carbon enrichment of the environment. The fact that USA has not agreed to Kyoto protocol is a reminder of how much the advanced countries can be stubborn on a commitment for carbon reduction.

For India, the Chinese model suits well. While the developmental agenda should not be compromised up on, India needs take measures to educate people and organizations on carbon economy, carbon recycling and nonconventional energy resources. Further, India needs effective and committed negotiators at the international fora at which the country's interests are protected. Not only should India take measures but also seem to be taking measures in good faith, as this will go a long way in convincing the international negotiators of the sincerity of efforts of the country for carbon reduction. Science, scientific pursuits and scientists should be encouraged to be proactive on environmental issues so that indictment of India by developed countries on issues such as Asian brown cloud and sea surface anoxia can be effectively countered with facts and figures. Oceans are most unexplored areas for understanding the basic CO_2 interactions. He complimented the timely organization of the programme and suggested that the proceedings should be brought out for its dissemination at the national level.

Dr. P.S. Yadav

Professor and Dean, Manipur University, Imphal

Forests play important role in climate change Mitigation through emission reduction, carbon sequestration and carbon substitution. Besides fixation of carbon dioxide in coal, soil and ocean, forests maintain high carbon stocks by reducing deforestation and forest degradation and promoting the sustainable managements of all types of forests. Thus forests play important role in climate change Mitigation through emission reduction, carbon sequestration and carbon substitution. The people will have to pay the price for emission of greenhouse gases in the near future and have to opt for green technology to mitigate the climate change.

44 scientists from different fields of specialization ranging from corporate, mining research institutes, universities have participated and delivered the talks on the theme of the programme. In the end, I extend my gratitude to Dr. Malti Goel for providing the platform to scientists from all over country to discuss the various aspects of Carbon capture and Carbon storage.

Dr. Shailja Sharma

Shell India

Global emissions are growing at a pace that is demanding attention towards all the options available.

- ☆ It is natural that governments should first seek to exhaust the most attractive technologies, especially Efficiency and Renewables, which pay for part of their own costs; CCS differs in that it offers only CO_2 reduction
- ☆ However, the Shell Energy Scenarios, as well as other studies, are suggesting that the most aggressive deployment of Efficiency and Renewables will still not be adequate to prevent emissions from exceeding the thresholds indicative of a 2°C increase in global mean temperature, in the context of global growth
- ☆ CCS technology demonstration has been seriously taken up and Shell is actively involved; we expect this process to accelerate
- ☆ Expansion of the Carbon market offers a route for lowering the cost of CCS
- ☆ It is also expected that CCS costs will be reduced upon the achievement of scale
- ☆ Financing arrangements that do not cannibalize CDM have been proposed for CCS, and should be considered and evaluated
- ☆ 'Satellite' (rather than global) agreements between interested parties could develop, hastening uptake
- ☆ The decision to undertake a CCS project is a sovereign decision that will be guided by considerations of implementability, including geological suitability, and risk perception
- ☆ However, it is in India's interests to develop the local know-how; therefore, setting up of a CCS Centre in India would be a step in the right direction
- ☆ A CCS Centre could bring together all of the different aspects of the CCS technology, policy and financing aspects; thereby create the full slate of capability that is required for it to become a resource for policy makers
- ☆ Under the PM's NAPCC, the Mission for Strategic Knowledge offers a framework within which to locate such a Centre

Mr. Ashish Sethia

Head of Research, New Energy Finance

- ☆ There are implications of new funding sources in a nascent industry like Carbon Capture and Storage. Demand potential for reduction of CO_2 by 2020 estimated as 2bt plus CO_2 per year and to be competitive, the carbon capture and storage should aim to deliver 10 per cent of reduction by 2020 *i.e.* 200 Mt of CO_2.
- ☆ European Union is ahead of USA in planning for number of CO_2 storage projects and out of the target of storage of 200Mt CO_2 injection, it is expected

to reach 108 Mt CO_2 with the share of EU being 42 Mt CO_2 in 2020, provided adequate funds are available.

- The finance required for 108 Mt CO_2/year will be \$80 billion. U.S. is moving ahead in pledging support for CCS projects, but oil companies and utilities are also required to provide support. As of now already there is a funding gap of \$56 billion and if the target of 200 Mt is to be achieved the funding gap widens to \$136 billion.
- Post combustion carbon capture and pre-combustion carbon capture technologies offer greater promise for deployment of CCS.

Some of these aspects are highlighted in the following presentation.

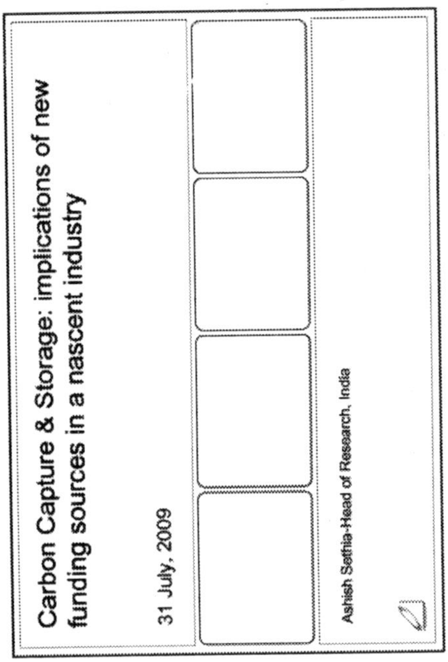
Carbon Capture & Storage: implications of new funding sources in a nascent industry
31 July, 2009
Ashish Sethia-Head of Research, India

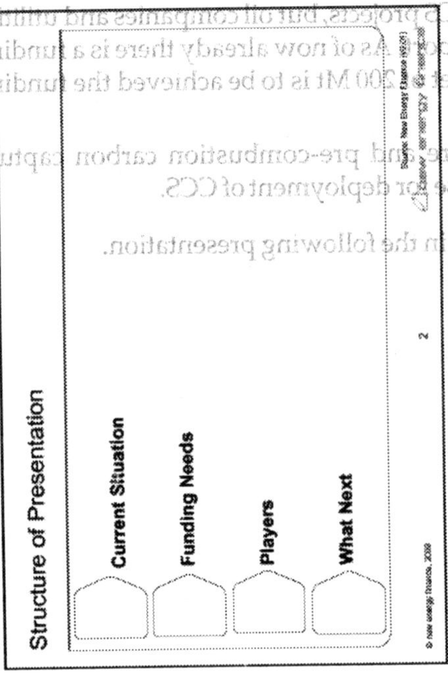
Structure of Presentation
Current Situation
Funding Needs
Players
What Next
2

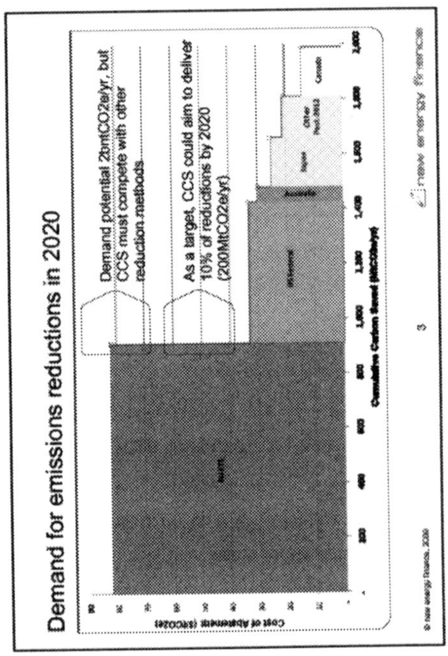
Demand for emissions reductions in 2020
Demand potential 2bntCO2e/yr, but CCS must compete with other reduction methods
As a target, CCS could aim to deliver 10% of reductions by 2020 (200MtCO2e/yr)
Cumulative Carbon Saved (MtCO2e/yr)
3
new energy finance

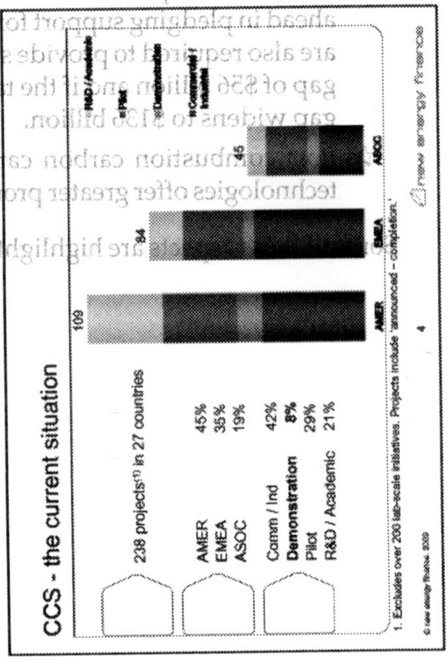
CCS - the current situation
238 projects(1) in 27 countries
AMER 45%
EMEA 35%
ASOC 19%
Comm / Ind 42%
Demonstration 8%
Pilot 29%
R&D / Academic 21%
109
84
45
AMER
EMEA
ASOC
1. Excludes over 200 lab-scale initiatives. Projects include 'announced – completion.'
4
new energy finance

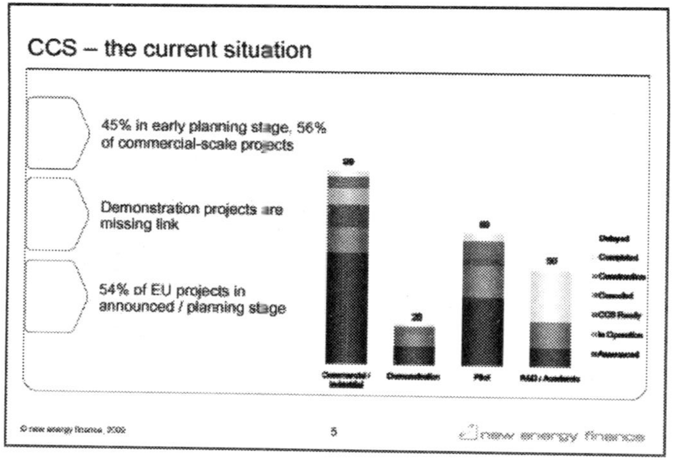
CCS – the current situation
45% in early planning stage, 56% of commercial-scale projects
Demonstration projects are missing link
54% of EU projects in announced / planning stage
© new energy finance, 2009
5
new energy finance

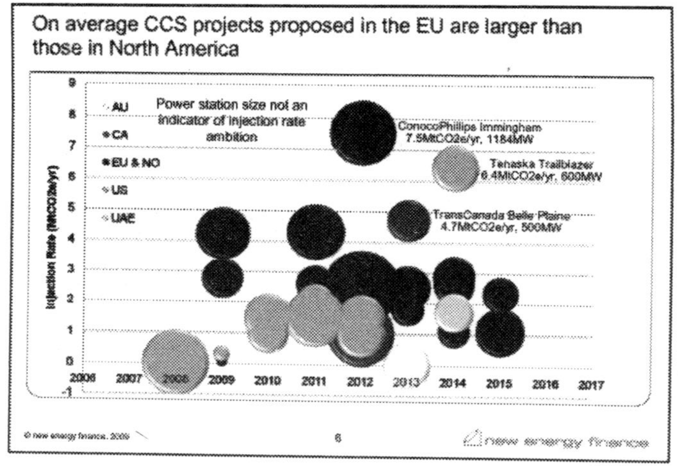
On average CCS projects proposed in the EU are larger than those in North America
Injection Rate (MtCO2e/yr)
AU
CA
EU & NO
US
UAE
Power station size not an indicator of injection rate ambition
ConocoPhillips Immingham 7.5MtCO2e/yr, 1184MW
Tenaska Trailblazer 6.4MtCO2e/yr, 600MW
TransCanada Belle Plaine 4.7MtCO2e/yr, 500MW
2006 2007 2008 2009 2010 2011 2012 2013 2014 2015 2016 2017
© new energy finance, 2009
6
new energy finance

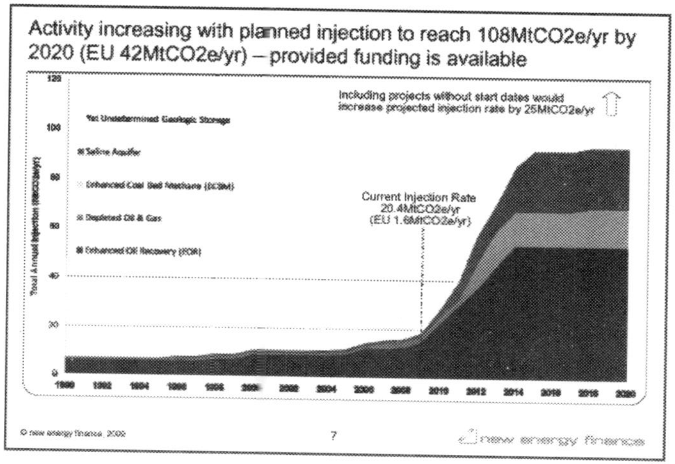
Activity increasing with planned injection to reach 108MtCO2e/yr by 2020 (EU 42MtCO2e/yr) – provided funding is available
Including projects without start dates would increase projected injection rate by 25MtCO2e/yr
Current Injection Rate 20.4MtCO2e/yr (EU 1.6MtCO2e/yr)
© new energy finance, 2009
7
new energy finance

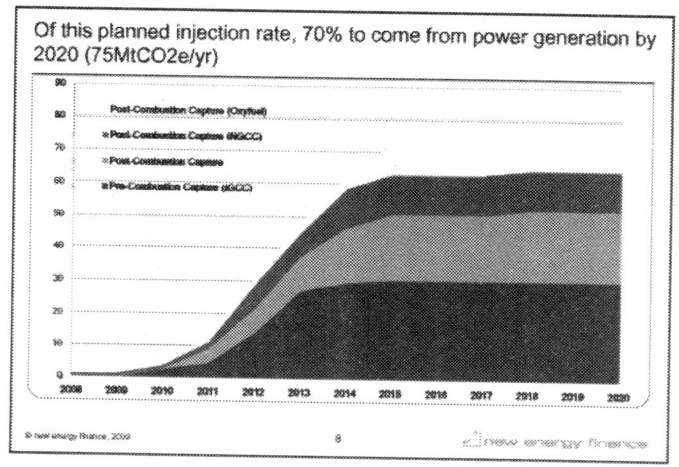
Of this planned injection rate, 70% to come from power generation by 2020 (75MtCO2e/yr)
2008 2009 2010 2011 2012 2013 2014 2015 2016 2017 2018 2019 2020
© new energy finance, 2009
8
new energy finance

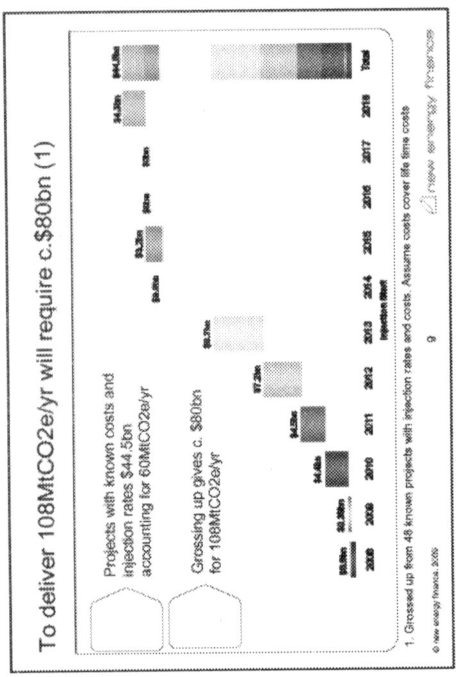
To deliver 108MtCO2e/yr will require c.$80bn (1)
Projects with known costs and injection rates $44.5bn accounting for 60MtCO2e/yr
Grossing up gives c. $80bn for 108MtCO2e/yr
1. Grossed up from 48 known projects with injection rates and costs. Assume costs cover life time costs
9

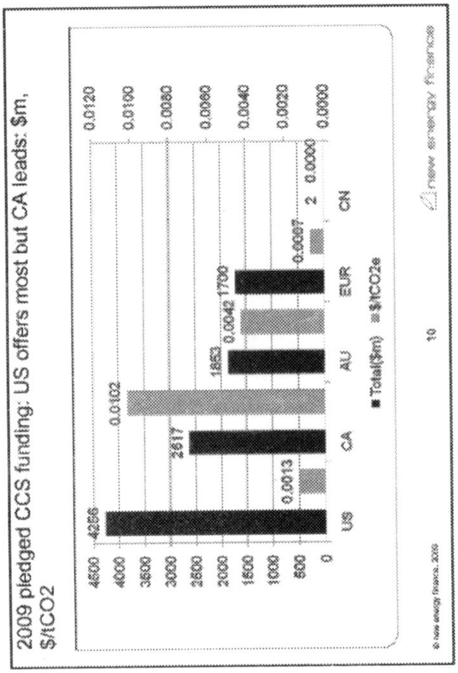
2009 pledged CCS funding: US offers most but CA leads: $m, $/tCO2
4256
0.0013
2617
0.0102
1853
0.0042
1700
0.0087
2
0.0000
US
CA
AU
EUR
CN
Total($m)
$/tCO2e
10

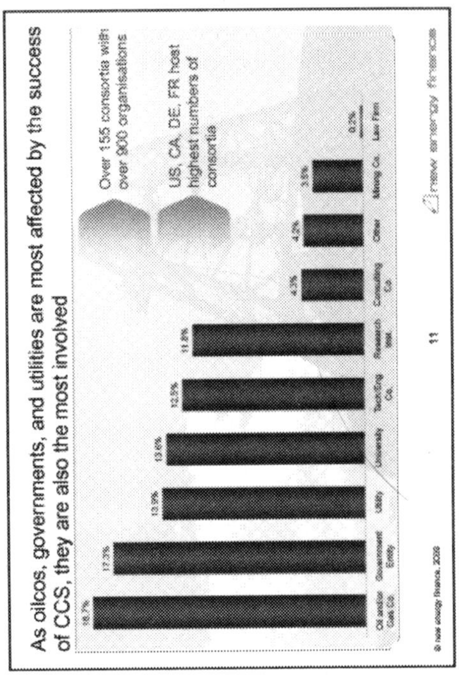
As oilcos, governments, and utilities are most affected by the success of CCS, they are also the most involved
Over 155 consortia with over 900 organisations
US, CA, DE, FR host highest numbers of consortia
11

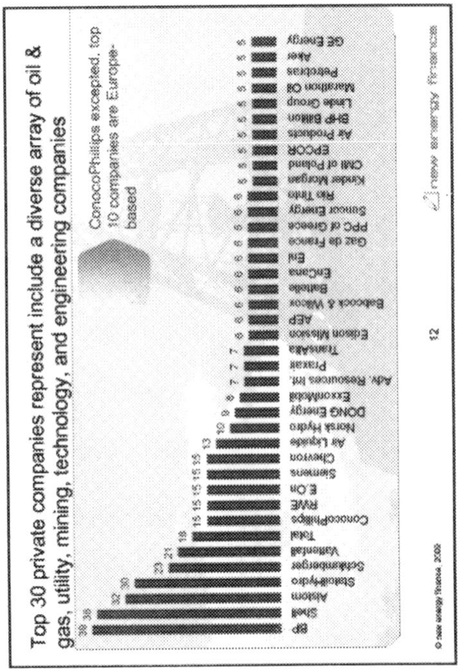
Top 30 private companies represent include a diverse array of oil & gas, utility, mining, technology, and engineering companies
ConocoPhillips excepted, top 10 companies are Europe-based
12

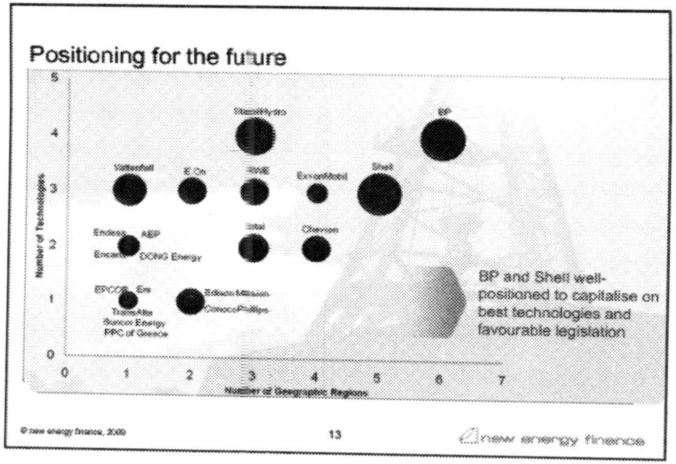
Positioning for the fu ure
Vattenfall
E.On
RWE
ExxonMobil
Shell
BP
Chevron
DONG Energy
BP and Shell well-positioned to capitalise on best technologies and favourable legislation
Number of Technologies
Number of Geographic Regions
0
1
2
3
4
5
6
7
© new energy finance, 2009
13
new energy finance

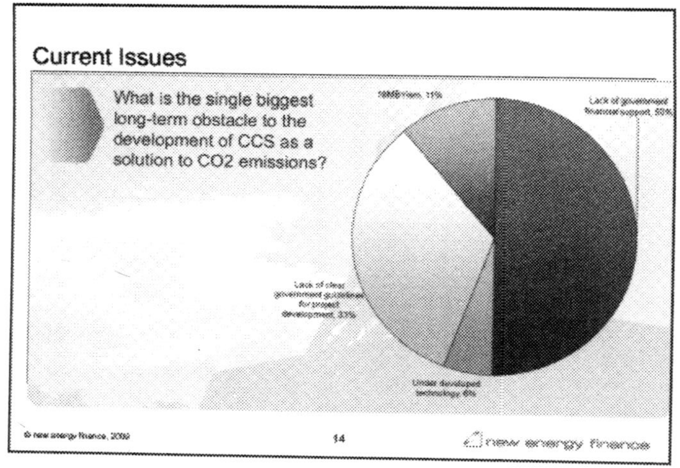
Current Issues
What is the single biggest long-term obstacle to the development of CCS as a solution to CO2 emissions?
© new energy finance, 2009
14
new energy finance

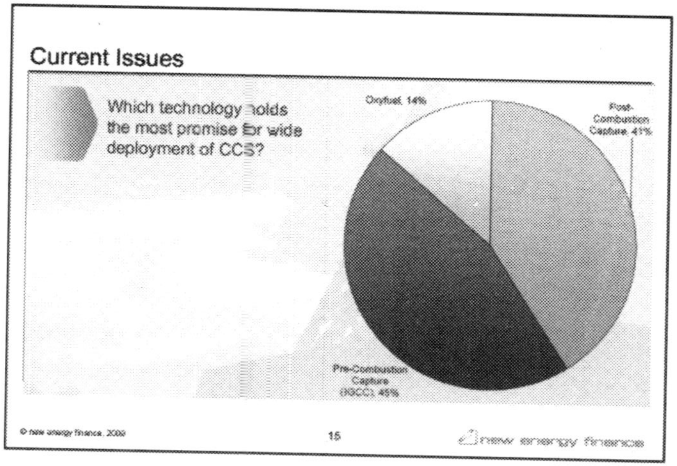
Current Issues
Which technology holds the most promise for wide deployment of CCS?
Oxyfuel, 14%
Post-Combustion Capture, 41%
Pre-Combustion Capture (IGCC), 45%
© new energy finance, 2009
15
new energy finance

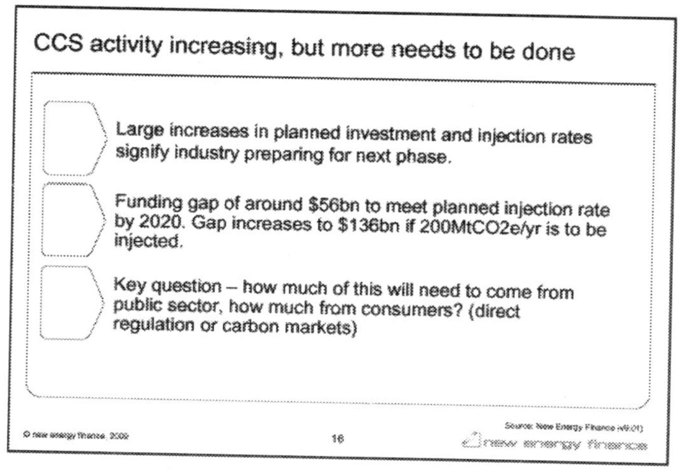
CCS activity increasing, but more needs to be done
Large increases in planned investment and injection rates signify industry preparing for next phase.
Funding gap of around $56bn to meet planned injection rate by 2020. Gap increases to $136bn if 200MtCO2e/yr is to be injected.
Key question – how much of this will need to come from public sector, how much from consumers? (direct regulation or carbon markets)
© new energy finance, 2009
16
new energy finance

Ms. Anitha Joy, Mr. Pradeep Kumar Pandey and Mr. Saket N. Niraj

National Hydroelectronic Power Corporation

HYDROPOWER: GHG AND ENVIRONMENTAL COSTS*

Hydropower contributes to about 19 per cent of global electricity production. It accounts for 24.7 per cent of the total installed capacity for power generation in India. Despite zero emissions during power generation, hydropower is not classified as a renewable energy source. Hydropower development is capital intensive, whereas operation and maintenance costs are low, thereby reducing the cost of electricity in the long run. Power generation do not involve effluent or solid waste generation, thereby keeping the environmental impacts low during operation.

In India hydropower development is invariably associated with stringent statutory regulations. Hydropower projects require payment of environmental costs under several heads before getting clearances for construction, especially if acquiring forest land is involved for project development. This write-up is an effort to bring into purview the GHG related issues of hydropower as well as the environmental costs paid up by hydropower developers in India.

The potential for hydropower generation is high in developing nations as compared to developed nations. The major environmental impact of hydropower development is involuntary displacement of population, which is predominantly rural in developing nations. The large-scale displacements that occurred due to dam building led to global campaigns against hydropower development. Such global campaigns against hydropower development included the accusation of GHG emissions from large reservoirs, albeit for a short time period.

Hydropower and GHG Emissions

The 1990s saw a global debate on GHG emissions from reservoirs of hydropower projects. Studies conducted by World Commission on Dams (WCD) and International Rivers Network (IRN) generated a controversy on the GHG emission potential of large dams, which emit gases such as methane and carbon dioxide due to submergence of biomass (forests). These studies threw a negative light on large dams as being potential emitters of GHGs. This argument was partially validated in a global forum, by the Intergovernmental Panel on Climate Change (IPCC), which in its draft of "IPCC guidelines for National GHG Inventories" included methods for estimation on GHG emissions from managed wetlands.

---

*The views expressed are not of the organization.

IPCC Draft Report

Volume 4 of the IPCC guidelines is on "Agriculture, Forestry and Other Land Use (AFOLU)". Chapter 7 of the AFOLU volume is on 'Wetlands'. The guidelines classified wetlands into two:

- ✩ Peat lands cleared and drained for production of peat for energy, horticultural and other uses and
- ✩ Flooded lands in reservoirs or impoundments for energy production, irrigation, navigation or recreation.

The section provided guidance for annual GHG inventories for emission of CO_2, CH_4 and N_2O from flooded lands. A three tiered approach was used in outlining methodologies that may be used for estimation. Tier 1 methods were designed to be the simplest to use, which could use globally available sources of activity data in case country-specific activity data were not available. Tier 2 methods were used for country or region specific data and Tier 3 used higher order methods including models and inventory measurement systems tailored to address national circumstances, repeated over time and driven by high-resolution activity data.

Default emission factors were given for various climatic zones to be used in the Tier 1 method. Tier 2 method for CO_2 had to use given country specific emissions. Tier 3 method was not given in detail. The default emission factors for Tier 1 method integrated some spatial (intra-reservoir and regional variations) and temporal variations (dry/rainy and other seasonal variations, inter-annual variations) in the emissions from reservoirs, as well as fluxes at the water-air interface of reservoirs.

The section on Lands converted to flooded lands gave methodology for emission estimates on lands such as forest land, cropland, grassland, settlements and other land converted to flooded land after removal of biomass before flooding. The method assumes that the carbon stock of the land prior to conversion is lost in the first year following conversion (decay is complete in one year).

The methods proposed in the inventory took into account the emissions either estimated on-ground or based on default emission factors, area of water body and the time period. The calculated emission is the gross emission from a reservoir and does not incorporate factors for natural pre-impoundment emissions or carbon contribution to reservoir from the catchment.

The final report "2006 IPCC Guidelines for National Greenhouse Gas Inventories" provided methodology for emission of CO_2 from Lands converted to flooded lands. Methodologies for estimation of CH_4 was restricted and for N_2O was excluded. Inclusion of reservoirs of energy generating units of hydropower posed the risk of hydropower being excluded from the category of developmental projects eligible for Certified Emission Reductions (CERs) under the Clean Development Mechanism.

Power Density Rule

The theory of GHG emissions from reservoirs was supported by the findings of a Thematic Review prepared as an input to the World commission on Dams. The thumb rule used is a recurrent yardstick applied in international discussion and referred in a paper by German Technical Co-operation (Deutsche Gesellschaft für Technische Zusammenarbeit or GTZ) regarding the power density of a project. Power density ratio is the ratio between power capacity/submergence area (W/m^2). An oft-cited rule of thumb is that *"If the power generated per unit area is less than 0.1 W/m^2, an efficient fossil-fuelled power station can be less damaging to the climate. If this figure is above 0.5 W/m^2, we may assume that the hydropower station has a lower greenhouse effect. This figure can be applied as a rough guide, particularly in tropical regions."* The calculated power density of a few projects studied by WCD is given below.

Table 1: Power Density and Flooded Area of a Few Reservoirs Studied for GHG Emissions

Plant	*Power (MW)*	*Flooded Area (km^2)*	*Power Density (W/m^2)*
Itaipu	12600	1350	9.33
Samuel	217	560	0.39
Tucurui	3960	2430	1.63
Balbina	250	2346	0.10
Churchill/Nelson River Development	1.824	1400	0.0012
Grand Rapids	0.192	1200	0.000114

Indian Scenario

In stark contrast to the above mentioned cases, the power density of many of the operating projects of NHPC is high as flooded area is quite low as given below.

Table 2: Power Density (W/m2) and Flooded Area in Some NHPC Owned Projects

Plant	*Power (MW)*	*Flooded Area (km^2)*	*Power Density (W/m^2)*
Chamera-II, HP	300	0.25	1200.00
Baira Siul, HP	180	0.152	1184.21
Dhauliganga (Uttaranchal)	280	0.28665	976.80
Teesta–V, Sikkim	510	0.6775	752.77
Rangit, Sikkim	60	0.129	465.12
Tanakpur, Uttaranchal	120	1.4	85.71
Salal-I and II, J&K	690	9.35	73.80
Chamera-I, HP	540	9.75	55.38
Loktak d/s, Manipur	90	2.57	35.02
Indirasagar, MP	1000	913.48	1.09

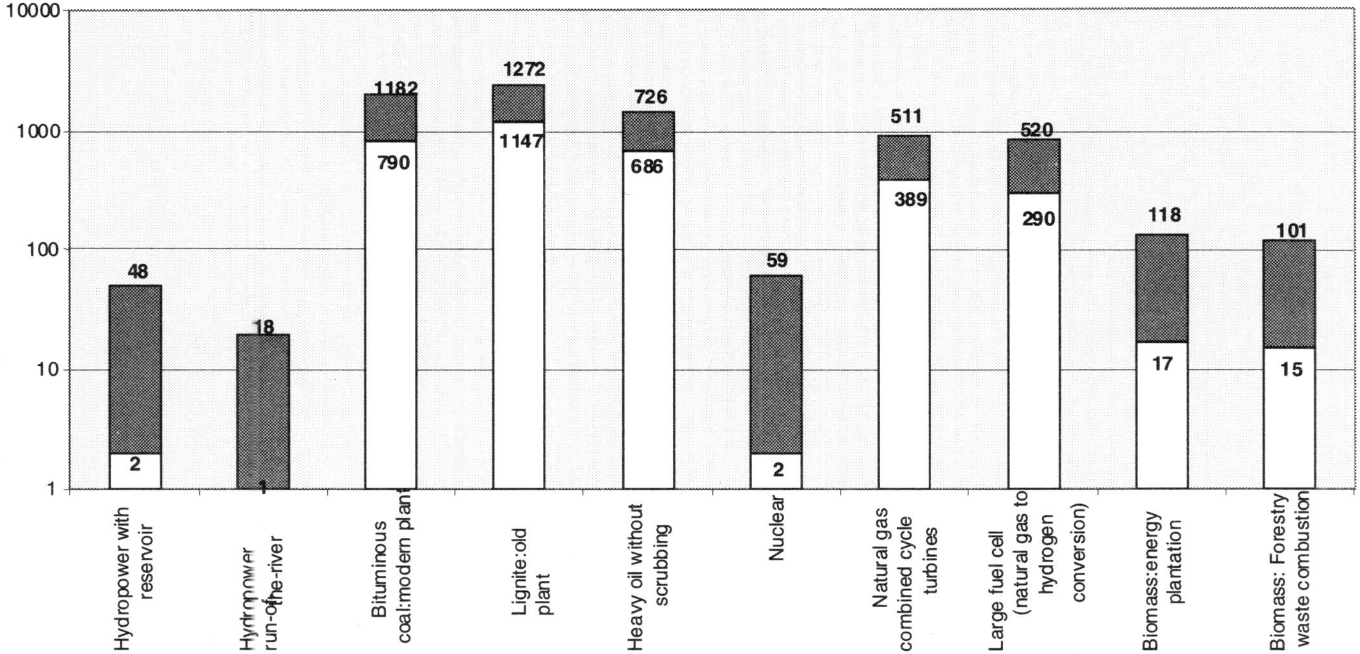

Figure 1: Comparative Chart Showing Range of Greenhouse Gas (kt eq. CO_2/TWh) Emitted from Various Sources of Electricity (IEA, May 2000)

Note: The shaded area represents the range of GHG Emissions from various sources of power generation.

The power densities of all projects except Indirasagar project are exceptionally high in comparison to the projects studied for GHG emissions, rendering the argument of GHG emissions from hydropower projects invalid in the Indian scenario. It may also be borne in mind that the reservoir of Indirasagar project is the largest in India, and the thumb rule used for the assessment of greenhouse effect of a reservoir does not hold good even for Indirasagar project.

A look at GHG emissions from different power sources in comparison to hydropower plants also indicates the low global warming potential of hydropower.

Avoided Carbon Emissions and Sequestered CO_2

Energy production by hydropower units now qualify for calculation of CERs, particularly for small hydropower projects. So far hydropower units with installed capacity up to about 500 MW have qualified for issuance of CERs under the Clean Development Mechanism. The Approved consolidated baseline and monitoring methodology ACM0002 of UNFCCC used for calculation of CERs from grid-connected renewable energy sources uses estimation of avoided carbon emissions for calculation of emission reduction units. A study that tried to analyse different scenarios (refer Table 2) for avoided carbon emission by hydroelectric plants found that hydropower is definitely a better option to control climate change.

Experience with Indian hydropower reservoirs reveals that although a large portion of forest land come under submergence, most often the whole stretch of land is not covered by forests. The land often includes waste land, shoals or river bed area and forests as well under the ownership of the Forest Department. Compensatory Afforestation in lieu of forest land diverted for non-forest purposes is mandatory in India for dams requiring/acquiring forestland. Most of the time the dam proponents afforest double the area of forestland acquired. Apart from this there is also plantation done under various management plans like landscaping, catchment area treatment, green belt plan, restoration of muck disposal and quarry sites, etc. All these plantation schemes start much before inundation, during the construction phase.

Table 2: Avoided Carbon Emissions (kg C/m^2) from Hydropower–Different Scenarios Using GWP from IPCC (1995)

Technology and Fuel/ Thermoelectric Plant	*Avoided Emissions (kg carbon/m ^{2}) by Hydroelectric Plants in Comparison to Fuel/Thermoelectric Plants*				
	P = 0.1	*P= 0.5*	*P = 1.0*	*P = 5.0*	*P = 10.0*
Conventional coal run	~28	24	89	609	1259
Conventional oil run	~40	9	59	450	960
Combined cycle/natural gas	~35	~13	14	234	509

Note: P denotes Power density ratio, i.e. W/m^2

Carbon Sequestration by Forestation

It is an established fact that growing forests is one means of sequestering carbon dioxide from the atmosphere as trees store carbon in their leaves, branches and roots. But it is not a fixed sequestration as eventually individual trees die and return that carbon to the atmosphere to be taken up by replacement trees. More and more area therefore must be turned into forests to store additional carbon. It is only through the growing of trees in un-forested or lightly forested area that forests can make any net gains in carbon sequestration and that is exactly what is being done in hydropower projects being executed in India. Planting trees in degraded forestland not only helps in habitat creation for flora and fauna but also helps in absorbing GHG emitted from various sources–natural/anthropogenic. Although the rate of carbon sequestration depends on various factors like plant species, soil characteristics, age, growth rate of the plant, mortality rate etc., studies carried out in various forests have revealed that 1 m^3 of wood captures about 1.3 t of CO_2. So annual sequestration in plantation of 20 m^3 of wood/ha is 26 t CO_2/ha. Moreover it was found out that the growth rate and the carbon sequestration potential of forests declines towards plant maturity. Studies have shown that in a matured forest, growth rate is largely offset by wood decay. Matured and climax forests are neither sources nor sinks of atmospheric carbon. In fact forests that experience a net loss of biomass through mortality due to disease or fire become net carbon emitters.

Environmental Costs of Hydropower Development

Hydropower development in India requires two major clearances from the environmental point of view–environmental clearance and forest clearance. For environmental clearance of a project, the project proponent is required to conduct an Environment Impact Assessment (EIA) and accordingly prepare an Environmental Management Plan (EMP) for the project taking into account the various environmental impacts of the project. The different management plans essentially include a compensatory afforestation plan, catchment area treatment plan, green belt plan, reservoir rim treatment plan, restoration plans for dumping and quarry sites, provision for free fuel for workers during construction, etc. Apart from these quite often due to remote locations of several projects, the project proponent takes care of the socioeconomic welfare and medical needs of the local population as well. Funds are earmarked under the EMP in the planning stage itself. The percentage of cost of EMP w.r.t the cost of project varies from 0.5 per cent to 4 per cent. The costs of other socio-economic welfare measures are recurrent and additional to the cost of EMP.

Net Present Value

Payment of Net Present Value (NPV) of forest land diverted for the project came into effect from 18th Sept 2003 following a Supreme Court verdict on the T N Godavarman Thirumulpad *vs* Union of India case. The case was a landmark case in the history of environmental legislation in India as it initiated an effort on monetarily quantifying the ecological cost of an area of forest land. In its judgements on the case in October 2002 and August 2003, the Hon'ble Supreme Court of India directed collection of Net Present Value of forests from user agencies. The court specified a range for the rates of foests from 5.80 lakhs per ha to 9.20 lakhs per ha, depending upon the quality of forest, density and type of species in the area. A recent decision on the same case has revised the range of rates to 4.38 per ha to 10.5 per ha.

The calculated NPV of forests diverted, as decided by the forest departments range from a few lakhs to a few hundred lakhs very often. The inclusion of cost of NPV along with other environmental costs often shoots up the environmental costs of a project.

Conclusion

From the above discussions it can be observed that hydropower is a low GHG emitting source of energy and at the same time in the Indian scenario it necessitates compensatory afforestation which in turn enables carbon sequestration as well. Besides, the environmental costs involved in hydropower development in India is also quite high

Main References

R.N Misra, General Manager (Planning); Devjani Patra, Environment Officer (Planning), National Hydroelectric Power Corporation Ltd; Hydropower and Global warming–the misunderstood science; India Hydro 2004, International Conference, New Delhi. www.indiahydro.org/inha_2004.

Dr. Ashutosh Mondal

CBM Cell, Mission– IIB, Geological Survey of India, Kolkata

CO_2 SEQUESTRATION AND ENHANCED COAL BED METHANE (ECBM) PRODUCTION: POSSIBILITIES IN INDIAN CONTEXT

CO_2 sequestration is a multifaceted aspect involving capture of carbon from atmosphere followed by transportation, injection into favorable sites and post-injection monitoring. The favorable sites for storage of CO_2 must be reliable in the sense that CO_2 will be stored there permanently at least for 1000 years and no leakage is preferable. In this backdrop, coal bed and saline aquifer are the most suitable storage sites where CO_2 will be fixed permanently by chemisorption and chemical reaction respectively. This write-up focuses on possibilities of CO_2 sequestration *vis-a-vis* enhanced coal bed methane production.

Coal being an adsorbent can trap gaseous sorbate like CO_2 and CH_4 within its micropores by chemisorption. Adsorption capacity of coal i.e abundance of micropores gradually increases with advancement of geochemical coalification process. Methane generated during thc coalification process is being trapped by the adsorbent (coal). Considering the methane adsorption capacity *vis-à-vis* methane generation capacity of a coal bed, it has been found that the coal beds with $R_{O\,max}$ up to 0.84 per cent are likely to be under-saturated whereas the coal beds with $R_{O\,max}$ greater than 0.84 per cent are likely to be over-saturated. With these backdrops, both the high rank as well as low rank coals are found suitable for CO_2 sequestration with or without enhanced coal bed methane production (ECBM).

Observations from Experiments at Laboratory and from Pilot Projects

Affinity of CO_2 is more than that of CH_4 on coal. Adsorption isotherm study of pure gaseous phases like CO_2 and CH_4 on coal absorbent reveals that the ratio of absorbed CO_2 and CH_4 generally ranges from 3: 1 to 4:1 depending on pressure. Injection of CO_2 into deep seated (>950m depth) brown coal with a permeability in the order of 100 md of San Juan basin in USA has improved methane recovery and it has been found that ratio of injected CO_2 to produced CH_4 is 3:1. Though the injectivity of CO_2 into the brown coal was reduced initially, may be due to swelling, it gradually increased. Thc pilot project operated in Yubari site, Ishkari Coalfield, Japan reveals that gas (CH_4) production increases with CO_2 injection and later it was dropped when injection was stopped, suggesting ECBM effect. In addition to that almost 90 per cent of injected CO_2 has been stored in the coal seam. The rate of injection was very slow initially but the gradual increase in injection rate has been observed later. In Silesian basin of Poland, CO_2 was injected at slow rate (200-800 m^3/day) into a highly undersaturated coal with permeability of around 1md. CO_2 breakthrough after small volume of injection indicates incomplete understanding of reservoir mechanics related to CO_2 injection. From the above observations it can be concluded that; a) experimental data on adsorption capacity of CO_2 and CH_4 is corroborated from pilot project, b) low permeability is one of the main controlling factors for slow rate of injection and c) highly undersaturated coal seams do not seem necessarily to be good candidates for CO_2 storage sites. It has been found that the permeability of coal reduced to less than 1 md below 1000m depth.

General Characteristics of Indian Coal

Coal seams are mostly banded in nature. Aerial continuity of coal seams is relatively small and seams are highly splitted. Coal characteristics vary widely both inter-basinally as well as intra-basinally. Vitrinite content is generally very low. Barring a few coalfields of Damodar Valley basin and parts of Sohagpur Coalfield and Satpura areas, coal seams are mostly of low rank and with high ash content. Moreover, cleats are mineralized at places.

Feasibility of CCS vis-a-vis ECBM

Coals occurring at depth of greater than 1km are not suitable for ECBM or CO_2 injection because of its very low permeability.

CBM potentiality is limited only to a few coalfields in India. Cost reduction of CCS project through ECBM is thus feasible to these coalfields only if other factors are favourable. In case of coal/lignite fields with highly undersaturated coal/lignite seam, cost reduction of CCS project through ECBM is untenable. Keeping in mind, the experience of Polish pilot project, the undersaturated nature of coal/lignite seams does not make these fields the automatic choices for CCS project.

Both CO_2 sequestration and/or ECBM by CO_2 injection into coal beds results in permanent wastage of coal resource, which must be taken into account. Moreover, if the CO_2 sequestration within the coal beds with or without ECBM becomes successful, CO_2 sequestration into coal beds will cover only a small fraction of the total budget of CO_2 reduction from the atmosphere.

Note: There are personal views of the Panelists and the editors, publishers and organizations are not responsible for the remarks.

Participants' Profile

Mr. Prashant is B.Tech. in Mining in 1996 and M.Tech in Open Cast Mining from Indian School of Mines, Dhanbad in 2000. Sri Prashant joined CIMFR in December 2001. He is extensively engaged in the area of mine stowing and backfilling, rock mechanics instrumentation for strata control, environment friendly disposal of coal ash in underground and surface mines and computer aided designing. He has received national recognition for Fly Ash Utilization by Ministry of Power, Ministry of Environment & Forest and Department of Science & Technology, Government of India, for R &D on the Use of Fly Ash for Stowing in Underground Mines. Sri Prashant is a life member of CAII and MGMI.

Mr. Shyam Nath Hazari is a Research fellow in the Methane Emission and Degasification Department at Central Institute of Mining and Fuel Research (CIMFR), Dhanbad. He has M.Sc. degree in Geology from Vinoba Bhave University, Hazaribagh. Mr. Hazari joined CIMFR in November 2008. Before joining CIMFR, he has worked as a Geologist in Total CBM Solutions, New Delhi from July 2007 to October 2008. He was the key person in the company to determine gas desorption parameters in respect of coal seams in Singarauli and Barmer-Sanchor basin. His research is primarily concerned with Coal Geology, Non-Conventional Energy and Remote Sensing application to Geosciences.

Mr. K. K. Singh is Technical Officer at Central Institute of Mining and Fuel Research (Erstwhile Central Mining Research Institute), Dhanbad. He has passed M.Sc. (First Class) in Chemistry and joined erstwhile CMRI in April 1999. He has conducted gas surveys in more than 50 underground coal mines to ensure safety against gas hazards. Mr. Singh has significant contributions to Coal Bed Methane Resource Assessment of different CBM blocks in India.

Mr. Anil Kumar Kar is Asst. Engineer, o/o- Engineer- in- Chief (Water Resources), Secha Sadan, Bhubaneswar, Pin-751001. He has done Civil Engineering from UCE, Burla, Orissa in 1989. Has completed M.Tech in Hydrology at I.I.T., Roorkee (2006-2008) and presently doing Ph.D in soft computing technique applied in flood forecasting. He has specialization in Stochastic hydrology, application of Artificial neural networks, Flood frequency analysis and Flood forecasting.

Mr. Narayan P. Gautam, (143/11 Shiva Panchayan Marg, Chhauni, Kathmandu, Nepal) did M.Sc. in Hydro-meteorology in 1998 from Tribhuvan University (TU), Kathmandu,Nepal (M.Sc. thesis is entitled, "mean annual and flood discharge analyses for Karnali, Narayani, Bagmati and Sapta Koshi river basins)". He is lecturer in the Department of Hydro-Meteorology, Tribhuvan University, Kathmandu, Nepal. Also Researcher at Total Quality Management (TQM) Nepal since April 2003 to date. He is life member, Society of Hydrologists and Meteorologists (SOHAM)-Nepal and is carrying out climate change related research as a M.Tech. final year student of Indian Institute of Technology, Roorkee.

Dr. Ashutosh Mondal, M.Sc in Applied Geology from Jadavpur University, Kolkata in the year 1987. He was engaged in CSIR sponsored Research Work on "Petrology of the Granulite facies rocks around Sunkarimetta, Andhra Pradesh" and obtained Ph.D from the same University in the year 1991. He was awarded "Prof. Nirmal Nath Chatterjee Medal" by the Asiatic Society for the year 1991 for making important contribution to the knowledge of economic geology. "National Mineral Award-2006" is conferred upon Dr. Mondal in recognition of his significant contribution in the field of Coal Discovery in the Sohagpur Coalfield. Presently he is associated in the field of Coal bed Methane and Carbon Sequestration.

Mr. Tarun Kr Mukherjee is Deputy Chief Mining Engineer at Coal India Ltd. He graduated in Mining Engineering from Indian School of Mines, Dhanbad in 1982 in 1st class with distinction. He has served Coal India Ltd (CIL) in different capacities in Eastern Coalfields Ltd (ECL), a Subsidiary of CIL. Since 2006, Sri Mukherjee has been working in Corporate Planning Division of CIL, which is responsible for planning CIL operations at a macro level. Sri Mukherjee is a life member of IE (India) and MGMI (India). He is also a member of MEAI. He has several papers to his credit. Sri Mukherjee has visited United Kingdom under Colombo Plan for attending a course on 'Underground Mine Mechanization' and France to attend an international seminar on CTL.

☐ **Mr. T. K. Chakraborty,** Chief Mining Engineer, Coal India Ltd., Kolkata.

☐ **Mr. R.S. Chakravarthy,** Chief of Corporate Communication and Asstt. General Manager (Marketing), MECON Ltd., Doranda, Vivekanand Path, Ranchi- 834002, Jharkhand.

☐ **Dr. Vikas Kumar,** Asstt. General Manager, MECON Ltd., Doranda, Vivekanand Path, Ranchi- 834002, Jharkhand.

☐ **Ms. Anitha Joy,** Environment Officer Planning Division, Corporate Office NHPC Ltd., Faridabad. She is M.Sc. (Environmental Studies), Cochin University of Science and Technology. She has research experience in the environmental factors that influence aquaculture production at Cochin University of Science and Technology. Presently she deals with the environmental aspects of hydropower development in the present capacity as Environment Officer.

☐ **Mr. Saket N. Niraj,** Asstt. Manager, R&D, NHPC Office Complex, Sector – 33, Faridabad – 121003 (Haryana).

☐ **Mr. Pradeep Kumar Pandey,** Asstt. Manager (Civil), Office of Executive Director – CBD, NHPC Office Complex, Sector – 33, Faridabad – 121003 (Haryana).

☐ **Dr. Manas Roychowdhury,** Senior Geologist, Geodata-I Division, Coal Wing, Geological Survey of India, Bhu-Bijnan Bhavan, Karunamoyee, Salt Lake City, Kolkata-91.

Units and Abbreviations

m	meter
mm	milli meter
ppm	parts per million
°C	degree centigrade
MW	mega watts
MU	million units
kWh	kilowatt hour
USD	US Dollar
%	per cent
W/m^2	watts per square meter
Mt	million tons
Gt	giga tons
K	degree Kelvin
g	gm
m^3	cubic meter
atm	Atmosphere
Btu	British thermal unit
BCM	billion cubic meter
TCM	trillion cubic meter
sq.km	square kilometer
l^{-1}	per litre
$t\,C\,ha^{-1}$	tons carbondioxide per hactare
yr	Year
EU	European Union
kcal	kilo calorie
GCV	gross calorific value